Transformational Plane Geometry

TEXTBOOKS in MATHEMATICS

Series Editors: Al Boggess and Ken Rosen

PUBLISHED TITLES CONTINUED

TEXTBOOKS in MATHEMATICS

Transformational Plane Geometry

Ronald N. Umble

Millersville University
Millersville, PA, USA

Zhigang Han

Millersville University
Millersville, PA, USA

CRC Press
Taylor & Francis Group
Boca Raton London New York

CRC Press is an imprint of the
Taylor & Francis Group, an **informa** business

A CHAPMAN & HALL BOOK

To our students —

whose curiosity and creativity are the inspiration for this book.

Contents

Foreword

by Maria Cristina Bucur and Lindsay Eisenhut

Graduate students in Mathematics Education
Millersville University of Pennsylvania

If you're asking yourself "What is transformational geometry?" or "Why am I required to study transformational geometry?" you are not alone. Not long ago, we found ourselves asking the same questions. As graduate students who had not taken a geometry course since high school, we initially felt intimidated by the material. However, through hard work, dedication, and open-mindedness, we came to understand and enjoy the content of this course.

Transformational geometry is a modern approach to Euclidean plane geometry, which, among other things, focuses on the study of rigid motions, i.e., translations, rotations, reflections, and glide reflections, and how they relate to congruence. Transformational geometry is a visual science that can be taught through engaging and creative pictorial exercises, and as you will see, transformational geometry is fun to learn and to teach.

At first, this course may seem more abstract than other geometry courses you may have had. Give yourself time to adjust to the structure and notation. If you commit yourself to reading the textbook carefully, understanding the proofs, and completing the exercises, the ideas will quickly come into focus. We hope that you enjoy this course as much as we did. We encourage you to ask questions whenever something is unclear to you. If teaching is your career goal, it is important to master the material and methodology, which is now the instructional standard for teaching geometry in secondary schools.

In the discussion that follows, we identify various ways this textbook relates to the national standards for teaching and learning mathematics. Additionally, we present suggestions for teaching these concepts to middle and high school students. We hope you find this information helpful as you begin your study of this exciting area of mathematics.

Transformational Geometry for Future Teachers: the Whys, the Dos, and the Don'ts

The remarks that follow make connections between *Transformational Plane Geometry* and two national standards for mathematics: National Council of Teachers of Mathematics' [NCTM] *Principles and Standards for School Mathematics* [PSSM] and the Common Core State Standards Initiative's *Common Core State Standards* [CCSS] *for Mathematics*. Additionally, we discuss some instructional best practices for teaching transformational geometry and identify some common pitfalls new teachers will want to avoid.

The adoption of national standards affects how mathematics, including geometry, is taught at the middle and high school levels. National standards aim to provide all students with a solid mathematical foundation by focusing on conceptual understanding, calculation speed and accuracy, and application of mathematical knowledge in real-world situations. National standards have a direct impact on instructional strategies as they dictate the curriculum, content order, and topic emphasis. Given the learning curve and period of adjustment that follows implementation of national standards, current and future educators alike need to be cognizant of the instructional implications.

I. The Relationship between Studying and Teaching Transformational Geometry

Current and future educators who expect to implement the national standards need to have a deep understanding of the concepts they will eventually teach. *Transformational Plane Geometry* helps to prepare future teachers in this regard, by motivating and developing the analytical thinking skills articulated in the NCTM's PSSM (2000) and the CCSS for mathematics (2009). These standards specify the mathematical reasoning, comprehension, and competences that students from prekindergarten through grade 12 need to achieve.

"As the standards have changed emphasis, I have attempted to shift my focus to what is current – however, I don't see that huge a difference in what the spirit of the documents said – instead, I see the current CCSS as 'fleshing out' what the intent was before – the exception being with the importance of proof."

Denise Young, Honors Geometry Teacher
Blue Valley West High
Overland Park, Kansas

Regarding geometry, the NCTM's PSSM directs students to investigate properties of geometric shapes, *"as well as to use visualization, spatial reasoning, and geometric modeling to solve problems"* (NCTM Executive Summary, 2000, p. 3). Additionally, students must learn to *"recognize reasoning and proof as fundamental aspects of mathematics"* (NCTM, 2014). *"Transformational geometry is one of the areas of geometry that has a potential to contribute to the development of students' reasoning and justification skills."* (Yanik, 2011, p. 231)

Transformational Plane Geometry encourages mathematical reasoning, by establishing theorems and propositions with proofs. And indeed, middle and high school students are required to give a written justification of their mathematical thinking. Brian Peltz, a high-school mathematics teacher at Interboro High School in Pennsylvania, has found that many of his students struggle with this standard: *"Writing a proof without any scaffolding was not going to happen at this level. I had some success with fill-in-the-blank proofs."* Although students need guidance when constructing proofs, teaching analytical thinking skills prepares them to apply logical reasoning in their future endeavors.

The content of *Transformational Plane Geometry* is related to many of the CCSS for mathematics. Consider the grade 8 Common Core geometry standard CCSS.MATH.CONTENT.8.G.A.4:

"Understand that a two-dimensional figure is similar to another if the second can be obtained from the first by a sequence of rotations, reflections, translations, and dilations; given two similar two-dimensional figures, describe a sequence that exhibits the similarity between them." (CCSSI, 2010, p. 55)

In order to meet this standard, teachers and their students must have a strong understanding of similarity and congruence, properties of rotations, reflections, translations, and dilations, and the concept of sequencing transformations.

These concepts are presented through exercises and examples in *Transformational Plane Geometry*. Some exercises require students to apply the material in real-world situations; others encourage students to construct theoretical arguments, which apply the definitions, theorems, and propositions presented in the text. Although university students study transformational geometry at a higher level than is presented in the secondary curriculum, the content is directly related to the content they will eventually teach.

II. Instructional Methodologies and Strategies

The constant evolution of secondary education mathematics standards requires teachers to regularly educate themselves on new instructional strategies. Three such strategies are:

1. "Let Students Build the Problem."(Meyer, 2010)

2. Incorporate Student Activities into Lesson Plans.

3. Take Advantage of All Available Resources.

"Let Students Build the Problem"

Albert Einstein once stated that, *"the formulation of a problem is often more essential than its solution, which may be merely a matter of mathematical or experimental skill"* (Meyer, 2010). In order to develop mathematical reasoning skills, students need to learn to construct and justify their own arguments, and not rely on a some specified procedure. One instructional methodology that promotes mathematical intuition asks students short questions that help them build the problem.

"...pedagogical choices made by the teacher, as manifested in the teacher's actions, are key to the type of classroom environment that is established and, hence, to students' opportunities to hone their proof and reasoning skills. More

specifically, the teacher's choice to pose open-ended tasks, ... engage in dialogue that places responsibility for reasoning on the students, analyze student arguments, and coach students as they reason, creates an environment in which participating students make conjectures, provide justifications, and build chains of reasoning." (Martin et al., 2005, p. 95).

In other words, posing open-ended questions and encouraging students to build the problem helps them develop math reasoning skills and provides them with a strong mathematical foundation.

Incorporate Student Activities into Lesson Plans

As students tend to disconnect during Socratic lessons, engaging in-class activities relating to real-world situations provide an ideal way for students to interact with and apply learned concepts. According to NCTM's PSSM, students should *"recognize and apply geometric ideas and relationships in areas outside the mathematics classrooms, such as art, science, and everyday life."* Denise Young, an honors geometry teacher at Blue Valley West High in Kansas, integrates this principle into her own teaching:

"Students were expected to complete two distinct art projects that [I] had them choose from: mandala (total symmetry), tessellations (all three isometries were available), miniature golf course hole showing the "hole in one" path (optimal path found by using a MIRA), quilt panel (all three isometries could be used), Islamic Tile (all three isometries can be found), Celtic Knot design (primarily rotational symmetry), Navajo Rug (all three isometries can be used)."

We find similar engaging exercises in *Transformational Plane Geometry.* For example, one of the book's activities requires students to classify frieze patterns. We found that such exercises not only relate geometric ideas to art, but they are also a fun way for students to apply the learned concepts.

From Exercise 7.4.5.

Take Advantage of All Available Resources

Given that students often require a deeper understanding than one single textbook can provide, teachers should utilize a variety of instructional resources including, but not limited to: their peers, multimedia resources, and additional textbooks. For example, when studying transformational geometry at Millersville University, *Transformational Plane Geometry* is accompanied by numerous hands-on activities developed by our instructor, constructions from Dayoub and Lott's MIRA-based exercise book *Geometry: Constructions and Transformations*, and several laboratory assignments adapted from Michael de Villiers' *Rethinking Proof with Geometer's Sketchpad*. Additionally, peer study-groups are an important and often underutilized resource. Although beginning teachers may feel intimidated by their more experienced colleagues, it is important to ask them questions, exchange ideas, and be willing to modify teaching methods.

While it is important to take advantage of all available resources, it is equally important to understand which teaching aids are appropriate and how to use them effectively. A group of geometry teachers we interviewed regularly use animations, software, MIRAs, worksheets, and wax paper to introduce transformations. When reinforcing the concepts, these teachers use protractors, rulers, MIRAs, hands-on activities, software, guided practice problems, daily homework assignments, quizzes, and group proof assessments. Malcolm Purvis, a high school mathematics teacher at Fallston High School in Maryland, stated that, *"On-line animations of reflections, rotations, glide reflections, and translations reinforce the concept[s] even more."*

 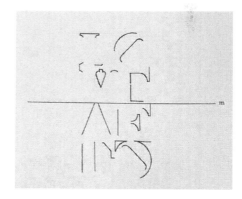

MIRA – geometric construction tool with the reflecting property of a mirror. From Dayoub and Lott, *Geometry: Constructions and Transformations*, Dale Seymour Pub., 1977.

III. Pitfalls and Common Problems Faced by New Teachers

Students preparing to teach mathematics need to be aware of some common pitfalls that can blindside them:

Don't Teach Too Much in One Day

In *The Don'ts and Don'ts of Teaching*, author Gary Rubinstein, a mathematics teacher in New York City, states that *"New teachers, particularly those without extensive student teaching,... create lessons that are too difficult, too long, or developmentally inappropriate"* (Rubinstein, 2012, p. 51). Although Rubinstein is an experienced teacher, he still struggles to avoid overpacking his lessons. It is important to give students enough time to understand, practice, and apply new concepts. When students are rushed, they easily become overwhelmed and miss the important conceptual ideas.

Don't Get Sidetracked by Teaching How to Use Technology

Lessons must always focus on learning and understanding mathematical concepts. Although teachers need to understand how to effectively apply instructional technology in the classroom, care must be taken to avoid focusing the lesson on how to use the technology. Technical malfunctions can distract from the lesson and cause students to become disengaged. According to the New York State Common Core Curriculum, *"technological tools are just tools, and their role in the classroom should be to facilitate learning, but not as an end in itself"* (Common Core, Inc., 2013, p. 17).

Furthermore, while dynamic geometric software, such as GeoGebra, Geometer's Sketchpad, and Live Geometry can be useful for demonstrating transformations, they may not be appropriate during a student's initial encounter with the material. When students are first exposed to rigid motions, manipulatives such as transparencies might better help students gain a tactile and visual familiarity with the concepts. Upon further exploration, *"students should be made aware of such software because such tools could be of later use, outside of the classroom"* (Common Core, Inc., 2013, p. 17).

Some instructional best practices used by the experienced teachers we interviewed are:

"[I] make sure that all of my students are on the same page [by reviewing concepts]."

Eliza Smith, High School Math Teacher
The Williams School
New London, Connecticut

"I try to emphasize the importance of always having a back-up plan for times when extra time remains in the class period."

Malcolm Purvis, Math Teacher
Fallston High School
Fallston, Maryland

"New teachers need to have a solid understanding of deductive reasoning in order to assist students with proof in geometry."

Rebecca Maryott, Honors Geometry Teacher
Manheim Township High School
Lancaster, Pennsylvania

"Don't be afraid to try things... Don't be driven by the book... Read the NCTM headlines each day to see what teachers in other parts of the country are doing to engage their students."

Denise Young, Honors Geometry Teacher
Blue Valley West High School
Overland Park, Kansas

Conclusion

In this discussion we presented evidence that the content in *Transformational Plane Geometry* relates directly to two national standards for mathematics–the NCTM's PSSM and the CCSS for mathematics. We identified some potential pitfalls that new teachers will want to avoid, as well as some instructional best practices used by experienced mathematics teachers.

Although it can be difficult for future teachers in training to connect the content in their required mathematics courses with the content they will eventually teach, our interviews with practicing professionals confirm that the deep conceptual understanding required for effective mathematics teaching is gained through the study of mathematics content courses at the university level. The study of transformational geometry, in particular, helps to develop students' mathematical intuition, logical reasoning, and justification skills.

Indeed, as you continue your studies in mathematics, your analytical thinking skills will continue to improve. As you enter the classroom and pass these skills along to your students, you will provide them with the tools they need to be successful in a wide variety of careers and professions. As stated by Dan Meyer, *"Math makes sense of the world. Math is the vocabulary for your own intuition"* (Meyer, 2010).

We wish you the best of luck in your mathematical studies. And as you will soon discover, transformational geometry is fun!

June 27, 2014

References

Common Core, Inc. (2013). New York State Common Core Mathematics Curriculum: Module 2. Retrieved from: //www.engageny.org/sites/default/files/resource/attachments/g8-m2-teacher_materials.

Common Core State Standards Initiative [CCSSI]. (2010). Common Core State Standards for Mathematics. Washington, DC: National Governors Association Center for Best Practices and the Council of Chief State School Officers.

Martin, T. S., McCrone, S. M. S., Bower, M. L. W., & Dindyal, J. (2005). The interplay of teacher and student actions in the teaching and learning of geometric proof. *Educational Studies in Mathematics*, 60(1), 95–124.

Meyer, D. (2010, March). Dan Meyer: Math class needs a makeover [Video file]. Retrieved from //www.ted.com/talks/dan_meyer_math_curriculum_makeover

National Council of Teachers of Mathematics. (2000). Executive Summary Principles and Standards for School Mathematics. Retrieved from: http://www.nctm.org/uploadedFiles/Math_Standards/12752_exec_pssm.pdf

National Council of Teachers of Mathematics. (2000). Principles and standards for school mathematics (Vol. 1).

National Council of Teachers of Mathematics. (2014). Reasoning and Proof Standard for Grades 6–8. Retrieved from: http://www.nctm.org/standards/content.aspx?id=26828

Rubinstein, G. (2012). The Don'ts and Don'ts of Teaching. Educational Leadership, 69(8), 50–52.

Yanik, H. B. (2011). Prospective middle school mathematics teachers preconceptions of geometric translations. *Educational Studies in Mathematics*, 78(2), 231–260.

Preface

This text is designed for a one-semester course in transformational plane geometry at the junior undergraduate level. The first two chapters serve as reference and provide the essential principles of classical Euclidean geometry we need. Consequently, we budget a limited amount of time to the content of these chapters in our classes. The main content begins with the definition of general transformations in Chapter 3; translations, rotations, and reflections are introduced in Chapter 4; and compositions of translations, rotations, and reflections are discussed in Chapter 5. The classification of isometries is presented in Chapter 6 and symmetry of plane figures is the focus of Chapter 7. The text concludes with a discussion of similarity in Chapter 8.

Although this text was written in response to the National Council of Teachers of Mathematics *Principles and Standards for School Mathematics* and the Common Core State Standards Initiative *Standards for Mathematical Practice*, which shift instructional focus from Euclid's purely axiomatic approach to Felix Klein's more visual and conceptual transformational point-of-view, this text addresses a general audience and offers every student of mathematics an opportunity to learn and appreciate plane geometry from a fresh hands-on interactive perspective.

Geometry is a visual science. Each concept needs to be contemplated in terms of some (often mental) picture. Consequently, a course in geometry needs to provide ample opportunity for students to create and (quite literally) manipulate pictures that represent the geometrical content of each concept. This text enables geometrical visualization in three ways: (1) Key concepts are motivated with exploratory activities using software specifically designed for performing geometrical constructions such as *Geometer's Sketchpad*. (2) Each new geometrical concept is introduced both synthetically (without coordinates) and analytically (with coordinates), and the exercises throughout the text accommodate both points of view. (3) The exercises at the end of each section include numerous geometric constructions, which utilize a reflecting instrument such as a MIRA.

Some experience with constructing proofs is assumed; however, the text is self-contained. Basic concepts from linear and abstract algebra are introduced gradually as they are needed. Prior knowledge of abstract algebra is not assumed.

Euclidean plane geometry is the study of size and shape of objects in the plane. It is one of the oldest branches of mathematics. Indeed, by 300 BC

Euclid had deductively derived the theorems of plane geometry from his five postulates. About 2000 years later in 1628, René Descartes introduced coordinates and revolutionized the discipline by using analytical tools to attack geometrical problems. To quote Descartes, "Any problem in geometry can easily be reduced to such terms that a knowledge of the lengths of certain straight lines is sufficient for its construction."

About 250 years later, in 1872, the great geometer Felix Klein capitalized on Descartes' analytical approach and inaugurated his famous *Erlangen Program*, which views plane geometry as the study of those properties of plane figures that remain invariant under the action of some group of transformations. Klein's startling observation that plane geometry can be completely understood from this point-of-view is the guiding principle of this text.

In this text we consider two families of transformations that act on the Euclidean plane $\mathbb{R}^2$ – *isometries,* which are length-preserving, and *similarities,* which are length-ratio-preserving, i.e., the ratio of the distance between two points in the image of a similarity to the distance between their preimages is constant. Isometries appear as reflections, translations, rotations, and glide reflections. Similarities include the isometries, which preserve the length-ratio 1, stretches, stretch reflections, and stretch rotations.

To the extent that congruence and similarity of plane figures can be understood in terms of isometries and similarities, this text provides a concrete visual alternative to Euclid's purely axiomatic approach to plane geometry. Indeed, as indicated in the National Council of Teachers of Mathematics: *Principles and Standards for School Mathematics* (http://www.nctm.org/standards/content.aspx?id=26835), the transformational approach is the methodology of choice for teaching geometry in secondary schools:

"In grades 9–12 all students should apply transformations and use symmetry to analyze mathematical situations:

- understand and represent translations, reflections, rotations, and dilations of objects in the plane by using sketches, coordinates, vectors, function notation, and matrices;

- use various representations to help understand the effects of simple transformations and their compositions."

A primary goal of this exposition is to identify and classify the isometries and similarities that act on the plane. Classification problems in mathematics are often profoundly difficult if not intractable, and the solution of a classification problem calls for great celebration! Indeed, the classification of plane isometries and similarities presented here is mathematics par excellence – a beautiful subtext of this course. The classification of isometries goes like this:

1. Every isometry is a composition of three or fewer reflections.

2. A composition of two reflections in parallel lines is a translation.

3. A composition of two reflections in intersecting lines is a rotation.

4. The identity is both a trivial translation and a trivial rotation.

5. Non-trivial translations are fixed point free, but fix every line in the direction of translation.

6. Non-trivial rotations fix exactly one point and are not translations.

7. A reflection fixes every point on its axis of reflection and is neither a translation nor a rotation.

8. A composition of three reflections in concurrent or mutually parallel lines is a reflection.

9. A composition of three reflections in non-concurrent and non-mutually parallel lines is a glide reflection.

10. A glide reflection is fixed point free and is neither a rotation nor a reflection.

11. A glide reflection fixes only its axis and is not a translation.

12. An isometry is exactly one of the following: a reflection, a rotation, a non-trivial translation, or a glide reflection.

The classification of isometries and similarities leads to the classification of several large families of plane figures up to symmetry type. In this text we define the symmetry type of a plane figure to be the "inner isomorphism class"of its symmetry group. An inner isomorphism of a symmetry group is the restriction of an inner automorphism of the group of all isometries to a subgroup of symmetries. Since inner isomorphisms preserve symmetry, two plane figures have the same symmetry type if and only if their symmetry groups are inner isomorphic. Thus the inner isomorphism class of a plane figure is a *perfect algebraic invariant*.

We wish to thank George E. Martin, author of the text *Transformation Geometry* (Springer-Verlag 1982), for his encouragement and permission to reproduce some of the exercises in his text. We wish to thank Millersville University graduate students Maria Cristina Bucur and Lindsay Eisenhut for contributing the Foreword, for helping us track down the quotes at the beginning of each chapter, for finding many of the graphical images and illustrations that appear throughout the text, and for helping us create the Hints and Answers to Selected Exercises in the Appendix. And finally, we wish to thank our former colleague Elizabeth Sell for carefully reading the manuscript and offering many helpful insights and suggestions.

October 21, 2014

Authors

Ronald N. Umble is a professor of mathematics at Millersville University of Pennsylvania, Millersville, Pennsylvania, where he has been a faculty member since 1984. Prior to this, he held faculty appointments at Hesston College, Hesston, Kansas (1975–84) and Blue Ridge Community College, Weyers Cave, Virginia (1974–75). He received his Ph.D. in algebraic topology under the supervision of James D. Stasheff from the University of North Carolina at Chapel Hill in 1983, his M.S. in mathematics from the University of Virginia in 1974, and his A.B. in mathematics from Temple University in 1972. He is a member of the Pennsylvania Zeta (Temple University) chapter of Pi Mu Epsilon and was the 1972 winner of the Zeta chapter's undergraduate paper competition. Ron loves teaching and research. He has directed numerous undergraduate research projects in mathematics; results from six of these projects were published. Ron is married to Dr. Diane Zimmerman Umble, Dean of the School of Humanities and Social Sciences at Millersville University. They reside in Lancaster, Pennsylvania, and have two adult children Kathryn and Eric.

Zhigang Han is an assistant professor of mathematics at Millersville University of Pennsylvania, Millersville, Pennsylvania, where he has been a faculty member since 2009. Prior to this, he was on the faculty at the University of Massachusetts Amherst (2006–09). He earned his Ph.D. in symplectic geometry and topology under the supervision of Dusa McDuff from Stony Brook University in 2006. Zhigang loves problem solving. He earned a perfect score in each of the 1992 National High School Mathematics Competition of China, the 1992 American High School Mathematics Examination, and the 1992 American Invitational Mathematics Examination. He lives in Lancaster, Pennsylvania, with his wife Yujen Shu and his daughter Erin Han.

Chapter 1

Axioms of Euclidean Plane Geometry

Euclid taught me that without assumptions there is no proof. Therefore, in any argument, examine the assumptions.

> Eric Temple Bell (1883–1960)
> Mathematician and science fiction author

Cogito ergo sum. (I think, therefore I am.)

> René Descartes (1596–1650)
> Father of analytic geometry

The first two chapters of this text develop the fundamental concepts and theorems of Euclidean plane geometry we need in our development of transformational plane geometry. We intentionally omit many important theorems of Euclidean plane geometry – including the famous Pythagorean Theorem – and our exposition freely applies many well-known fundamental facts from other branches of mathematics.

There are many different axiomatic systems from which to choose. Some authors begin with the minimum number of independent postulates needed for a rigorous development of geometry. Others begin with more postulates than necessary and allow some redundancy in exchange for shorter proofs of technical results. Since we wish to begin our exposition of transformational geometry as soon as possible, we make the latter choice here.

In the development of any area of mathematics, new terms are always defined in terms of existing ones. However, to avoid circularity among the definitions, certain primitive notions are left undefined. Such notions are called *undefined terms*. Our exposition assumes five undefined terms, namely, *point, line, distance, half-plane,* and *angle measurement.* This chapter introduces ten *postulates*, which spell out the properties these undefined terms possess.

1.1   The Existence and Incidence Postulates

In this section we introduce the Existence and Incidence Postulates, which specify the properties of the undefined terms *point* and *line*.

The Existence Postulate. There are at least two points.

The Incidence Postulate. Every line is a set of points. Through every two distinct points A and B, there is exactly one line l, denoted by $l = \overleftrightarrow{AB}$ (see Figure 1.1).

$$A \qquad\qquad\qquad\qquad B$$

Figure 1.1. The unique line determined by points A and B.

Definition 1 *The set of all points is called the **plane**.*

Definition 2 *Two lines l and m are **parallel**, written $l \parallel m$, if $l = m$ or $l \cap m = \varnothing$. If l and m are not parallel, we write $l \nparallel m$.*

Although some authors do not consider equal lines to be parallel, we do so here because this fits with our exposition of transformational plane geometry to follow.

Theorem 3 *Two nonparallel lines intersect at exactly one point.*

Proof. Let l and m be two lines such that $l \nparallel m$. By negating the definition of parallels, $l \neq m$ and $l \cap m \neq \varnothing$. Since $l \cap m \neq \varnothing$, there is a point $P \in l \cap m$. To show that P is unique, assume there is a point $Q \neq P$ such that $Q \in l \cap m$. Since P and Q are on l, and P and Q are on m, $l = m$ by the Incidence Postulate. But $l \neq m$, so this is a contradiction. Therefore l and m intersect at exactly one point. ∎

Exercises

1. Prove that there exists a line.

1.2 The Distance and Ruler Postulates

In this section we introduce the Distance and Ruler Postulates, which spell out the properties of the undefined term *distance*.

The Distance Postulate. Given points A and B, there is a unique real number, denoted by AB, called the distance from A to B.

Definition 4 *Let A and B be sets and let $f : A \to B$ be a function. Then f is **injective** (or **one-to-one**) if distinct elements of A take distinct values in B, i.e., if $x \neq y$ in A, then $f(x) \neq f(y)$ in B. The function f is a **surjective** (or **onto**) if the range of f is B, i.e., given any point $y \in B$, there exists an element $x \in A$ such that $f(x) = y$. If f is both injective and surjective, it is **bijective**.*

It is common to use the word *injection* to mean injective function, *surjection* for surjective function, and *bijection* for bijective function.

The Ruler Postulate. Given a line l, there is a bijection $f : l \to \mathbb{R}$ such that if A and B are points on l, then $AB = |f(A) - f(B)|$.

Theorem 5 below lists the properties of the distance function. These properties explain why we refer to AB as the distance from A and B.

Theorem 5 (Properties of the Distance Function) *Let A and B be points. The distance from A to B satisfies the following properties:*

1. *(Symmetry) $AB = BA$,*

2. *(Non-negativity) $AB \geq 0$, and*

3. *(Non-degeneracy) $AB = 0$ if and only if $A = B$.*

Proof. Let A and B be points. First observe that some line l contains A and B. If $A \neq B$, there is the line $l = \overleftrightarrow{AB}$ by the Incidence Postulate. If $A = B$, there is a point $C \neq A$ by the Existence Postulate, and there is the line $l = \overleftrightarrow{AC}$ by the Incidence Postulate. In either case, A and B are on l.

Let $f : l \to \mathbb{R}$ be a bijection given by the Ruler Postulate. Since A and B are on l, by algebra $AB = |f(A) - f(B)| = |f(B) - f(A)| = BA$, which proves symmetry (1), and also by algebra, $AB = |f(A) - f(B)| \geq 0$, which proves non-negativity (2).

For non-degeneracy (3), first assume $A = B$. Then $f(A) = f(B)$ since f is a function, and $AB = |f(A) - f(B)| = 0$. Conversely, if $AB = 0$, then $|f(A) - f(B)| = 0$ and by algebra, $f(A) = f(B)$. Since f is a bijection, $A = B$.
∎

The bijection $f : l \to \mathbb{R}$ in the Ruler Postulate is called a *coordinate function* of l since it maps each point A on l to its *coordinate* $f(A)$ on the real number line $\mathbb{R}$. Using this language, the Ruler Postulate can be restated as follows: *For every line there exists a coordinate function.* Note however, that such coordinate functions are not unique (see Exercise 1).

If A and B are distinct points, there is a particularly useful coordinate function that assigns the coordinate 0 to A and a positive coordinate to B.

Theorem 6 (The Ruler Placement Postulate) *For every pair of distinct points A and B, there is a coordinate function $f : \overleftrightarrow{AB} \to \mathbb{R}$ such that $f(A) = 0$ and $f(B) > 0$.*

Proof. Let A and B be distinct points and consider the line $l = \overleftrightarrow{AB}$ given by the Incidence Postulate. By the Ruler Postulate, there exists a coordinate function $g : l \to \mathbb{R}$. Let $c = -g(A)$ and define $h : l \to \mathbb{R}$ by $h(X) = g(X) + c$ for every point X on l. Then h is a coordinate function by Exercise 1 and $h(A) = g(A) + c = g(A) - g(A) = 0$. Since h is a bijection, $h(B) \neq 0$. Thus either $h(B) < 0$ or $h(B) > 0$. If $h(B) > 0$, set $f = h$; otherwise define $f : l \to \mathbb{R}$ by $f(X) = -h(X)$. Then f is a coordinate function by Exercise 1. Since $f(A) = -h(A) = 0$ and $f(B) = -h(B) > 0$, the proof is complete. ∎

Definition 7 *Three points A, B, and C on the same line are **collinear**. Otherwise A, B, and C are **non-collinear**.*

Definition 8 *Let A, B, and C be distinct points. Point C is **between** A and B, written $A - C - B$, if A, B, and C are collinear and $AC + CB = AB$.*

Proposition 9 *Let A, B, and C be distinct points. Then $A - C - B$ if and only if $B - C - A$.*

Proof. Assume that $A - C - B$. By definition, A, B, and C are collinear and $AB = AC + CB$. Since $AC = CA$, $CB = BC$, and $AB = BA$, we have $BA = AB = AC + CB = CA + BC = BC + CA$ by substitution. Therefore $B - C - A$. The proof of the converse is similar and is left to the reader. ∎

Definition 10 *The set*

$$\overline{AB} = \{A, B\} \cup \{P \mid A - P - B\}$$

*is called the **segment** from A to B. The points A and B are the **endpoints** of $\overline{AB}$; all other points of $\overline{AB}$ are **interior points** of $\overline{AB}$. The set*

$$\overrightarrow{AB} = \overline{AB} \cup \{P \mid A - B - P\}$$

*is called the **ray** from A through B. The point A is the **endpoint** of $\overrightarrow{AB}$.*

Definition 11 *Two segments $\overline{AB}$ and $\overline{CD}$ are **congruent**, written $\overline{AB} \cong \overline{CD}$, if $AB = CD$.*

In the remainder of this section, we will establish some results that follow from the Distance and Ruler Postulates.

Theorem 12 (Betweenness Theorem for Points) *Let $A, B,$ and C be distinct points on line l, and let $f : l \to \mathbb{R}$ be a coordinate function for l. Then $A - C - B$ if and only if $f(C)$ is between $f(A)$ and $f(B)$.*

Proof. Let $A, B,$ and C be distinct points on line l, and let $f : l \to \mathbb{R}$ be a coordinate function for l.

($\Rightarrow$) Assume $A - C - B$. Then $AC + CB = AB$. By the Ruler Postulate, $f(A), f(B),$ and $f(C)$ are distinct, and

$$|f(A) - f(C)| + |f(C) - f(B)| = |f(A) - f(B)|.$$

Let $x = f(A) - f(C) \neq 0$ and $y = f(C) - f(B) \neq 0$. Then $x + y = f(A) - f(B)$, and $|x| + |y| = |x + y|$. By algebra, x and y have the same sign. But if x and y are positive, then $f(A) > f(C) > f(B)$; and if x and y are negative, then $f(A) < f(C) < f(B)$. In either case $f(C)$ is between $f(A)$ and $f(B)$.

($\Leftarrow$) Assume that $f(C)$ is between $f(A)$ and $f(B)$. Then either $f(A) < f(C) < f(B)$ or $f(A) > f(C) > f(B)$. But in either case,

$$|f(A) - f(C)| + |f(C) - f(B)| = |f(A) - f(B)|.$$

Therefore $AC + CB = AB$ by the Ruler Postulate, and $A - C - B$ by definition. ∎

Corollary 13 *Given three distinct collinear points, exactly one of them is between the other two.*

Proof. See Exercise 4. ∎

Definition 14 *Let A and B be distinct points. A point M is a **midpoint** of $\overline{AB}$ if $A - M - B$ and $AM = MB$.*

Theorem 15 (Existence and Uniqueness of Midpoints) *Let A and B be distinct points. Then $\overline{AB}$ has a unique midpoint.*

Proof. Let A and B be distinct points. We first prove that there exists a point M such that M is a midpoint of $\overline{AB}$. Consider the unique line $l = \overleftrightarrow{AB}$ given by the Incidence Postulate. By the Ruler Placement Postulate, there is a coordinate function $f : l \to \mathbb{R}$ such that $f(A) = 0$ and $f(B) = b > 0$. Since f is surjective, there exists a point M on l such that $f(M) = \frac{b}{2}$. By the Ruler Postulate, $AB = |f(A) - f(B)| = |0 - b| = b$, $AM = |f(A) - f(M)| = |0 - \frac{b}{2}| = \frac{b}{2}$, and $MB = |f(M) - f(B)| = |\frac{b}{2} - b| = \frac{b}{2}$. Hence $AM + MB = AB$ and $AM = MB$. By definition, M is a midpoint of $\overline{AB}$. This proves existence. We leave the proof of uniqueness as an exercise for the reader. ∎

Exercises

1. Let $f : l \to \mathbb{R}$ be a coordinate function for line l. Prove that

   (a) For every constant $c \in \mathbb{R}$, the function $g : l \to \mathbb{R}$ defined by $g(X) = f(X) + c$ is also a coordinate function for l.
   (b) The function $h : l \to \mathbb{R}$ defined by $h(X) = -f(X)$ is also a coordinate function for l.

2. Let A, B, and C be three distinct points such that $\overrightarrow{AB} = \overrightarrow{AC}$. Prove that either $A - B - C$ or $A - C - B$.

3. Let A, B, and C be three distinct points such that C lies on ray $\overrightarrow{AB}$. Prove that $A - B - C$ if and only if $AB < AC$.

4. Prove Corollary 13.

5. Prove the uniqueness of midpoints (Theorem 15).

1.3 The Plane Separation Postulate

In this section, we introduce the Plane Separation Postulate, which states that the plane is divided by a line into two disjoint convex half-planes. Note that *half-plane* is one of the undefined terms.

Definition 16 *A set of points S is **convex** if $\overline{AB} \subseteq S$ for all $A, B \in S$.*

The Plane Separation Postulate. Give a line l, the points off l form two disjoint nonempty sets H_1 and H_2, called the half-planes bounded by l, with the following properties:

1. H_1 and H_2 are convex sets.

2. If $P \in H_1$ and $Q \in H_2$, then $\overline{PQ}$ cuts l (see Figure 1.2).

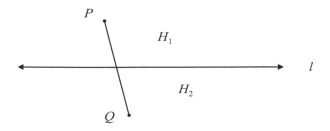

Figure 1.2. The Plane Separation Postulate.

Two points A and B off line l are **on the same side** of l if they lie in the same half-plane bounded by l; otherwise A and B are **on opposite sides** of l. The following proposition follows immediately from the Plane Separation Postulate.

Proposition 17 *Let A and B be points off line l. Then A and B are on the same side of l if and only if $\overline{AB} \cap l = \varnothing$; they are on opposite sides of l if and only if $\overline{AB} \cap l \neq \varnothing$.*

In the remainder of this section, we define angles, triangles, and related notions.

Definition 18 *An **angle** is the union of two rays $\overrightarrow{AB}$ and $\overrightarrow{AC}$ such that A is not between B and C. The angle is denoted by either $\angle BAC$ or $\angle CAB$. Point A is the **vertex** of the angle; rays $\overrightarrow{AB}$ and $\overrightarrow{AC}$ are the **sides** of the angle.*

Definition 19 *Let $\angle BAC$ be an angle. If $\overrightarrow{AB} \neq \overrightarrow{AC}$, then the **interior** of $\angle BAC$ is the intersection of the half-plane containing B and bounded by $\overleftrightarrow{AC}$ and the half-plane containing C and bounded by $\overleftrightarrow{AB}$ (see Figure 1.3). If $\overrightarrow{AB} = \overrightarrow{AC}$, then the **interior** of $\angle BAC$ is defined to be the empty set.*

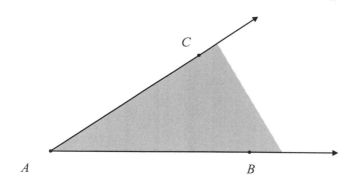

Figure 1.3. The interior of $\angle BAC$ is shaded.

Definition 20 *Let $A, B,$ and C be non-collinear points. The **triangle** $\triangle ABC$ is the union of segments $\overline{AB}, \overline{BC},$ and $\overline{AC}$. The points $A, B,$ and C are the **vertices** of the triangle and the segments $\overline{AB}, \overline{BC},$ and $\overline{AC}$ are the **sides** of the triangle.*

Theorem 21 (Pasch's Axiom) *Let $\triangle ABC$ be a triangle and let l be a line that misses the vertices of $\triangle ABC$. If l cuts $\overline{AB}$, then l cuts exactly one of $\overline{AC}$ or $\overline{BC}$.*

Proof. Let $\triangle ABC$ be a triangle and let l be a line that cuts $\overline{AB}$ but misses the vertices of $\triangle ABC$. Then by Proposition 17, A and B are on opposite sides of l. By the Plane Separation Postulate, each of the two disjoint half-planes bounded by l contains exactly one of the points A or B. Since C is off l, it lies in exactly one of these two half-planes. Thus C is on the same side of l as A or B, but not both. If C and A are on the same side, then C and B are on opposite sides, and by Proposition 17, l cuts $\overline{BC}$ but not $\overline{AC}$. On the other hand, if C and B are on the same side, then C and A are on opposite sides, and by Proposition 17, l cuts $\overline{AC}$ but not $\overline{BC}$. In either case, l cuts exactly one of $\overline{AC}$ or $\overline{BC}$. ∎

Exercises

1. True or false. Prove the statement that is true and disprove the statement that is false.

   (a) The intersection of two convex sets is convex.

   (b) The union of two convex sets is convex.

2. Prove that the interior of every angle is convex.

3. Let l be a line, A a point on l, and B a point not on l. Prove that all points on the ray $\overrightarrow{AB}$ except the endpoint A are on the same side of l.

4. Let $\triangle ABC$ be a triangle. Prove that one of the half-planes bounded by $\overleftrightarrow{BC}$ contains no points on $\triangle ABC$.

1.4　The Protractor Postulate

In this section, we introduce the Protractor Postulate, which is a collection of several postulates that specify the properties of the undefined term *angle measurement*. Notice that the undefined term here is angle measurement – *not* the angle.

The Protractor Postulate. For every angle $\angle BAC$ there is a real number $\mu(\angle BAC)$, called the *measure* of $\angle BAC$, with the following properties:

1. **(The Angle Measurement Postulate)** For every angle $\angle BAC$, $0° \leq \mu(\angle BAC) < 180°$, and $\mu(\angle BAC) = 0°$ if and only if $\overrightarrow{AB} = \overrightarrow{AC}$.

2. **(The Angle Construction Postulate)** If $0 < r < 180$ and H is a half-plane bounded by $\overleftrightarrow{AB}$, there is a unique ray $\overrightarrow{AE}$ such that $E \in H$ and $\mu(\angle BAE) = r°$.

3. **(The Angle Addition Postulate)** If D is a point in the interior of $\angle BAC$, then $\mu(\angle BAD) + \mu(\angle DAC) = \mu(\angle BAC)$ (see Figure 1.4).

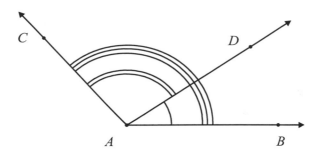

Figure 1.4. The Angle Addition Postulate.

Recall that the Ruler Postulate asserts the existence of bijection from a line l to $\mathbb{R}$. Similarly, the Protractor Postulate asserts the existence of a bijection from the set of all angles with side $\overrightarrow{AB}$ and interiors in a half-plane bounded by $\overleftrightarrow{AB}$ to the interval $(0, 180)$.

Definition 22 *Two angles $\angle BAC$ and $\angle EDF$ are **congruent**, written $\angle BAC \cong \angle EDF$, if $\mu(\angle BAC) = \mu(\angle EDF)$.*

Definition 23 *The angle $\angle BAC$ is **right** if $\mu(\angle BAC) = 90°$, is **acute** if $0° < \mu(\angle BAC) < 90°$, and is **obtuse** if $\mu(\angle BAC) > 90°$.*

The Continuity Postulate and the Supplement Postulate, which are stated below, actually follow from the Protractor Postulate. However, as mentioned earlier, we treat them as additional postulates to avoid their lengthy technical proofs and move as quickly as possible to the primary objective of this exposition.

The Continuity Postulate. Let A, B, C, and D be points such that C and D lie on the same side of $\overleftrightarrow{AB}$ as in Figure 1.5. Then D is in the interior of $\angle BAC$ if and only if $\mu(\angle BAD) < \mu(\angle BAC)$. Furthermore, if D is in the interior of $\angle BAC$, then the ray $\overrightarrow{AD}$ cuts the interior of segment $\overline{BC}$.

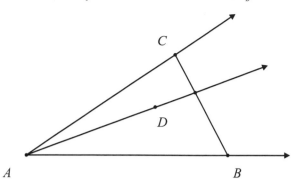

Figure 1.5. The Continuity Postulate.

We need the following two definitions before we can state the Supplement Postulate.

Definition 24 *Two angles ∠BAD and ∠DAC form a **linear pair** if B − A − C (see Figure 1.6).*

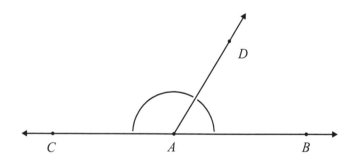

Figure 1.6. Angles ∠BAD and ∠CAD form a linear pair.

Definition 25 *Two angles ∠BAC and ∠EDF are **supplementary** if $\mu(\angle BAC) + \mu(\angle EDF) = 180°$.*

The Supplement Postulate. Two angles that form a linear pair are supplementary.

In the remainder of this section, we introduce the notions of angle bisector, perpendicular bisector, and vertical angles.

Definition 26 *Let A, B, and C be non-collinear points as in Figure 1.7. An **angle bisector** of ∠BAC is a ray $\overrightarrow{AD}$ such that*

1. *D is in the interior of ∠BAC, and*

2. $\mu(\angle BAD) = \mu(\angle DAC)$.

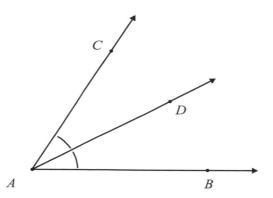

Figure 1.7. The angle bisector $\overrightarrow{AD}$.

Theorem 27 (Existence and Uniqueness of Angle Bisectors) *Let A, B, and C be non-collinear points. Then $\angle BAC$ has a unique angle bisector.*

Proof. Let A, B, and C be non-collinear points.

<u>Existence</u>: By the Angle Measurement Postulate, $0° < \mu(\angle BAC) < 180°$. Thus $0° < \frac{1}{2}\mu(\angle BAC) < 90°$. By the Angle Construction Postulate, there exists a unique ray $\overrightarrow{AD}$ such that D and C are on the same side of $\overleftrightarrow{AB}$ and $\mu(\angle BAD) = \frac{1}{2}\mu(\angle BAC)$. Note that $\mu(\angle BAD) < \mu(\angle BAC)$. By the Continuity Postulate, D is in the interior of $\angle BAC$. By the Angle Addition Postulate, $\mu(\angle BAD) + \mu(\angle DAC) = \mu(\angle BAC)$. By algebra, we have $\mu(\angle DAC) = \frac{1}{2}\mu(\angle BAC) = \mu(\angle BAD)$. By definition, $\overrightarrow{AD}$ is an angle bisector of $\angle BAC$. This proves existence.

<u>Uniqueness</u>: Suppose $\overrightarrow{AE}$ is another angle bisector for $\angle BAC$. By definition, E is in the interior of $\angle BAC$ and $\mu(\angle BAE) = \mu(\angle EAC)$. By the Angle Addition Postulate, $\mu(\angle BAE) + \mu(\angle EAC) = \mu(\angle BAC)$. By algebra, we have $\mu(\angle BAE) = \frac{1}{2}\mu(\angle BAC)$. By substitution, $\mu(\angle BAE) = \mu(\angle BAD)$. Since E is in the interior of $\angle BAC$, E and C lie on the same side of $\overleftrightarrow{AB}$. Thus E and D lie on the same side of $\overleftrightarrow{AB}$. By the uniqueness part of the Angle Addition Postulate, $\overrightarrow{AE} = \overrightarrow{AD}$. This proves uniqueness. ∎

Definition 28 *Two lines l and m are **perpendicular**, denoted by $l \perp m$, if they form right angles.*

By the Supplement Postulate and simple algebra, two perpendicular lines form four right angles.

Proposition 29 *Let l be a line and let P be a point on l. Then there exists a unique line m through P and perpendicular to l.*

Proof. Let l be a line and let P be a point on l. For existence, choose a point Q on l distinct from P. By the Angle Construction Postulate, there exists a ray $\overrightarrow{PR}$ such that $\mu(\angle QPR) = 90°$. Let $m = \overleftrightarrow{PR}$; then P lies on m and $m \perp l$. This proves existence. Uniqueness follows from uniqueness in the Angle Construction Postulate and the proof is complete. ∎

Definition 30 *Let A and B be distinct points as in Figure 1.8. A **perpendicular bisector** of $\overline{AB}$ is a line l such that*

  1. *the midpoint M of $\overline{AB}$ is on l, and*

  2. *$\overleftrightarrow{AB} \perp l$.*

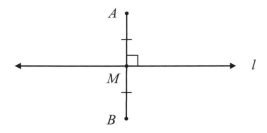

Figure 1.8. The perpendicular bisector of $\overline{AB}$.

Theorem 31 (Existence and Uniqueness of Perpendicular Bisectors) *Let A and B be distinct points. Then $\overline{AB}$ has a unique perpendicular bisector.*

Proof. See Exercise 1. ∎

Definition 32 *Angles $\angle BAC$ and $\angle DAE$ are **vertical angles** if either $B - A - D$ and $C - A - E$, or $B - A - E$ and $C - A - D$ (see Figure 1.9).*

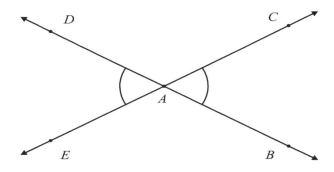

Figure 1.9. Vertical angles $\angle BAC$ and $\angle DAE$.

Theorem 33 (Vertical Angles Theorem) *Vertical angles are congruent.*

Proof. See Exercise 2. ∎

Exercises

1. Prove Theorem 31 (Existence and Uniqueness of Perpendicular Bisectors).

2. Prove Theorem 33 (Vertical Angles Theorem).

1.5   The Side-Angle-Side Postulate and the Euclidean Parallel Postulate

In this section, we introduce the Side-Angle-Side Postulate and the Euclidean Parallel Postulate.

Definition 34 *Triangles* $\triangle ABC$ *and* $\triangle DEF$ *are* **congruent**, *written* $\triangle ABC \cong \triangle DEF$, *if* $\overline{AB} \cong \overline{DE}, \overline{BC} \cong \overline{EF}, \overline{AC} \cong \overline{DF}, \angle BAC \cong \angle EDF, \angle ABC \cong \angle DEF$, *and* $\angle ACB \cong \angle DFE$.

In other words, two triangles are congruent if all three pairs of corresponding sides are congruent and all three pairs of corresponding angles are congruent.

The Side-Angle-Side Postulate (SAS). Let $\triangle ABC$ and $\triangle DEF$ be triangles. If $\overline{AB} \cong \overline{DE}$, $\angle ABC \cong \angle DEF$, and $\overline{BC} \cong \overline{EF}$, then $\triangle ABC \cong \triangle DEF$ (see Figure 1.10).

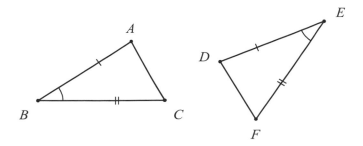

Figure 1.10. The Side-Angle-Side Postulate.

The Side-Angle-Side Postulate states that if two pairs of corresponding sides and the pair of corresponding angles between them are congruent, all six congruence conditions in Definition 34 are satisfied. In the next chapter, we will use the SAS Postulate to prove other triangle congruence theorems, including ASA, AAS, SSS, and Hypotenuse-Leg (HL). For now we will illustrate an application of the SAS Postulate in the proof of the following proposition, which will be useful in our study of transformational plane geometry.

Proposition 35 *Let l be a line, let P be a point on l, and let A and A' be points off l. If l is the perpendicular bisector of $\overline{AA'}$, then P is equidistant from A and A', i.e., $PA = PA'$.*

Proof. Let l be a line, let P be a point on l, and let A and A' be points off l such that l is the perpendicular bisector of $\overline{AA'}$. Let M be the midpoint of $\overline{AA'}$. We proceed by cases.

<u>Case 1</u>: $P = M$. Then $PA = PA'$ by definition of midpoint.

<u>Case 2</u>: $P \neq M$. Since M is the midpoint of $\overline{AA'}$, $AM = A'M$ so that $\overline{AM} \cong \overline{A'M}$. Also $\angle PMA \cong \angle PMA'$ since both are right angles, and $\overline{PM} \cong \overline{PM}$. By SAS, $\triangle PMA \cong \triangle PMA'$. Thus by the definition of congruent triangles, $\overline{PA} \cong \overline{PA'}$, and equivalently, $PA = PA'$.

In either case, $PA = PA'$ and the proof is complete. ∎

The definition of congruent triangles asserts that *Corresponding Parts of Congruent Triangles are Congruent*. This is commonly abbreviated as "CPCTC." We could use this to shorten the last sentence in the proof of Case 2 above like this: Thus by CPCTC, $\overline{PA} \cong \overline{PA'}$, and equivalently, $PA = PA'$.

Our final postulate is the Euclidean Parallel Postulate.

The Euclidean Parallel Postulate. For every line l and for every point P off l, there exists exactly one line m such that P is on m and $m \parallel l$.

Many statements are equivalent to the Euclidean Parallel Postulate. One such statement of particular historical importance is Euclid's Postulate V from Book I of his celebrated work *Elements*. The wording in the statement of Euclid's Postulate V below is written in today's language and differs somewhat from the original. But first we need the following definition:

Definition 36 *Let l and m be lines. A **transversal** is a third line t that cuts l and m at two distinct points.*

Euclid's Postulate V. If two lines l and m are cut by a transversal t such that the sum of the measures of the two interior angles on one side of t is less than $180°$, then l and m intersect on that side of t.

Exercises

1. Let A, B, and C be non-collinear points such that $AB = AC$. Prove that if D is on the angle bisector of $\angle BAC$, then $BD = CD$.

2. Prove that if segments $\overline{AA'}$ and $\overline{BB'}$ intersect at a common midpoint M, then $AB = A'B'$.

3. Given lines l, m, and n such that $l \parallel m$ and $l \parallel n$, is it possible that m and n intersect at some point? Explain your reasonings.

4. Determine whether the following analog of the Euclidean Parallel Postulate is true or false. Explain your reasoning.

   For every line l and for every point P on l, there exists exactly one line m such that P is on m and $m \parallel l$.

Chapter 2

Theorems of Euclidean Plane Geometry

The study of geometry contributes to helping students develop the skills of visualization, critical thinking, intuition, perspective, problem-solving, conjecturing, deductive reasoning, logical argument and proof.

Keith Jones, Math. Educator
University of Southampton

In this chapter we establish some essential theorems of Euclidean Plane Geometry by building upon the undefined terms *point, line, distance, half-plane,* and *angle measurement,* and the following postulates introduced in Chapter 1:

1. The Existence Postulate

2. The Incidence Postulate

3. The Distance Postulate

4. The Ruler Postulate

5. The Plane Separation Postulate

6. The Protractor Postulate

7. The Continuity Postulate

8. The Supplement Postulate

9. The Side-Angle-Side Postulate

10. The Euclidean Parallel Postulate

As mentioned perviously, our goal is to develop the machinery we need in our exposition of transformational plane geometry. Consequently, the discussion in this chapter is limited to those theorems from Euclidean plane geometry that will be applied.

2.1   The Exterior Angle Theorem

Before we can state our first theorem in this section, we need the following definition:

Definition 37 *Let* $\triangle ABC$ *be a triangle. The angles* $\angle CAB$, $\angle ABC$, *and* $\angle BCA$ *are* **interior angles** *of the triangle. An* **exterior angle** *of the triangle is an angle that forms a linear pair with one of the interior angles (see Figure 2.1).*

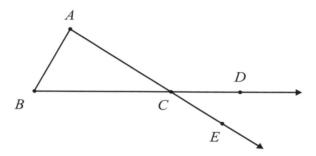

Figure 2.1. Exterior angles $\angle ACD$ and $\angle BCE$.

Theorem 38 (Exterior Angle Theorem) *The measure of an exterior angle of a triangle is greater than the measure of either non-adjacent interior angle.*

Proof. Let $\triangle ABC$ be a triangle and let D be a point on $\overleftrightarrow{BC}$ such that $B - C - D$. We must prove that $\mu(\angle ACD) > \mu(\angle BAC)$ and $\mu(\angle ACD) > \mu(\angle ABC)$. Let M be the midpoint of $\overline{AC}$. By the Ruler Postulate, there is a point F on $\overrightarrow{BM}$ such that $\overline{BM} \cong \overline{MF}$. By the Vertical Angles Theorem, $\angle AMB \cong \angle CMF$. By SAS, $\triangle AMB \cong \triangle CMF$. Thus $\angle BAM \cong \angle FCM$, or equivalently, $\mu(\angle BAC) = \mu(\angle FCA)$ (see Figure 2.2).

Now F and B are on opposite sides of $\overleftrightarrow{AC}$ and so are D and B. Therefore F and D are on the same side of $\overleftrightarrow{AC}$ by the Plane Separation Postulate. Furthermore, since $\overline{AM}$ misses line $\overleftrightarrow{CD}$, A and M are on the same side of $\overleftrightarrow{CD}$ by Proposition 17, and F and M are on the same side of $\overleftrightarrow{CD}$ for the same reason. Therefore F and A are on the same side of $\overleftrightarrow{CD}$ by the Plane Separation Postulate. By definition, F is in the interior of $\angle ACD$. Hence by the Continuity Postulate, $\mu(\angle ACD) > \mu(\angle FCA)$. Combining this with the previously established equality $\mu(\angle FCA) = \mu(\angle BAC)$ we conclude that $\mu(\angle ACD) > \mu(\angle BAC)$.

To prove that $\mu(\angle ACD) > \mu(\angle ABC)$, choose a point E on $\overleftrightarrow{AC}$ such that $A - C - E$. Then $\mu(\angle BCE) > \mu(\angle ABC)$ by the argument given above. Finally, $\mu(\angle ACD) = \mu(\angle BCE)$ by the Vertical Angles Theorem. Therefore $\mu(\angle ACD) > \mu(\angle ABC)$ and the proof is complete. ∎

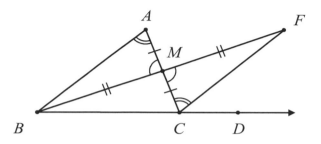

Figure 2.2. Proof of the Exterior Angle Theorem.

Given a point P and a line l, our next theorem asserts that there exists a unique line through P perpendicular to line l. Note that Proposition 29 has already established this fact when P is on l. While the proof of Proposition 29 is a simple application of the Angle Construction Postulate, Theorem 39 establishes a fact essential to define a reflection in transformational plane geometry, namely, if l is a line and P is a point off l, the image of P under the reflection in l is the unique point Q such that l is the perpendicular bisector of $\overline{PQ}$. The point Q is constructed in the proof of existence below.

Theorem 39 (Existence and Uniqueness of Perpendiculars) *Let l be a line and let P be a point. Then there exists a unique line m through P and perpendicular to l.*

Proof. Let l be a line and let P be a point. The case with P on l was established in Proposition 29, so assume that P is off l.

Existence: By the Ruler Postulate, there exist two distinct points A and B on l. By the Protractor Postulate, there exists a ray $\overrightarrow{AR}$ on the opposite side of l from P such that $\angle BAP \cong \angle BAR$. By the Ruler Postulate, we can choose a point Q on $\overrightarrow{AR}$ such that $\overline{AP} \cong \overline{AQ}$ (see Figure 2.3).

Let $m = \overleftrightarrow{PQ}$. It suffices to show that $m \perp l$. By the Plane Separation Postulate, segment $\overline{PQ}$ cuts line l at some point F. If $F = A$, then $\angle BFP$ and $\angle BFQ$ are supplements. Since they are also congruent, both are right angles and $m \perp l$. If $F \neq A$, then $\triangle FAP \cong \triangle FAQ$ by SAS. Therefore $\angle PFA \cong \angle QFA$ by CPCTC. Since $\angle PFA$ and $\angle QFA$ are also supplements, both are right angles and $m \perp l$. This completes the proof of existence.

Uniqueness: Suppose that there exists a line $m' \neq m$ such that P is on m' and $m' \perp l$. Let F' be the point at which l cuts m'. Then $F' \neq F$ by Theorem 3. Note that $\triangle PFF'$ has an exterior angle at F' and an interior angle at F

both measuring 90°. Since this contradicts the Exterior Angle Theorem, the proof of uniqueness is complete. ∎

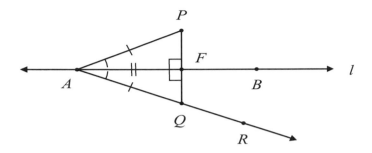

Figure 2.3. Proof of existence of perpendiculars.

For future reference, the line m is called the **perpendicular** from P to l, and the point F at which m cuts l is called the **foot** of the perpendicular from P to l.

Exercises

1. Prove that at least two interior angles of a triangle must be acute.

2.2   Triangle Congruence Theorems

Recall Definition 34, which states that two triangles are congruent if all three pairs of corresponding sides are congruent and all three pairs of corresponding angles are congruent. The Side-Angle-Side Postulate states that if two pairs of corresponding sides and the pair of corresponding angles between them are congruent, all six congruence conditions in Definition 34 are satisfied. In this section, we will develop four additional triangle congruence theorems, namely, ASA, AAS, SSS, and the Hypotenuse-Leg Theorem (HL).

Theorem 40 (ASA) *If $\triangle ABC$ and $\triangle DEF$ are triangles such that $\angle CAB \cong \angle FDE$, $\overline{AB} \cong \overline{DE}$, and $\angle ABC \cong \angle DEF$, then $\triangle ABC \cong \triangle DEF$.*

Proof. Let $\triangle ABC$ and $\triangle DEF$ be two triangles such that $\angle CAB \cong \angle FDE$, $\overline{AB} \cong \overline{DE}$, and $\angle ABC \cong \angle DEF$. We must prove that $\triangle ABC \cong \triangle DEF$.

By the Ruler Postulate, there exists a point C' on the ray $\overrightarrow{AC}$ such that $\overline{AC'} \cong \overline{DF}$. By SAS, $\triangle ABC' \cong \triangle DEF$ (see Figure 2.4). Then $\angle ABC' \cong \angle DEF$ by CPCTC. Thus $\angle ABC \cong \angle ABC'$. By the uniqueness part of Angle Construction Postulate, $\overrightarrow{BC} = \overrightarrow{BC'}$. By Theorem 3, $\overrightarrow{BC}$ cuts $\overleftrightarrow{AC}$ at most at one point. Therefore $C = C'$ and $\triangle ABC \cong \triangle DEF$. ∎

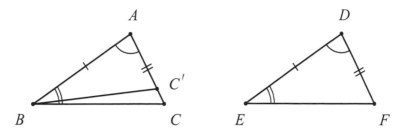

Figure 2.4. Proof of ASA.

Definition 41 *A triangle with two congruent sides is* **isosceles**; *a triangle with no two sides congruent is* **scalene**.

Theorem 42 (Isosceles Triangle Theorem) *Let* $\triangle ABC$ *be a triangle. Then* $\angle ABC \cong \angle ACB$ *if and only if* $\overline{AB} \cong \overline{AC}$.

Proof. Let $\triangle ABC$ be a triangle.
($\Leftarrow$) Assume that $\overline{AB} \cong \overline{AC}$. Since $\overline{AB} \cong \overline{AC}$, $\angle BAC \cong \angle CAB$, and $\overline{AC} \cong \overline{AB}$, we have $\triangle ABC \cong \triangle ACB$ by SAS and $\angle ABC \cong \angle ACB$ by CPCTC.
($\Rightarrow$) Assume that $\angle ABC \cong \angle ACB$. Since $\angle ABC \cong \angle ACB$, $\overline{BC} \cong \overline{CB}$, $\angle ACB \cong \angle ABC$, we have $\triangle ABC \cong \triangle ACB$ by ASA and $\overline{AB} \cong \overline{AC}$ by CPCTC. ∎

Theorem 43 (Scalene Inequality) *Let* $\triangle ABC$ *be a triangle. Then* $AB > AC$ *if and only if* $\mu(\angle ACB) > \mu(\angle ABC)$.

Proof. Let $\triangle ABC$ be a triangle.
($\Rightarrow$) Assume that $AB > AC$. By the Ruler Postulate, there exists a point B' such that $A - B' - B$ and $AB' = AC$ (see Figure 2.5). Note that $\mu(\angle ACB) > \mu(\angle ACB')$ by the Angle Addition Postulate, $\mu(\angle ACB') = \mu(\angle AB'C)$ by the Isosceles Triangle Theorem, and $\mu(\angle AB'C) > \mu(\angle ABC)$ by the Exterior Angle Theorem. Thus

$$\mu(\angle ACB) > \mu(\angle ACB') = \mu(\angle AB'C) > \mu(\angle ABC).$$

That is, $\mu(\angle ACB) > \mu(\angle ABC)$.

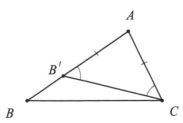

Figure 2.5. $AB > AC$ implies $\mu(\angle ACB) > \mu(\angle ABC)$.

($\Leftarrow$) We prove the contrapositive. Assume that $AB \le AC$. If $AB = AC$, then $\mu(\angle ACB) = \mu(\angle ABC)$ by the Isosceles Triangle Theorem; if $AB < AC$, then $\mu(\angle ACB) < \mu(\angle ABC)$ by the first part of the theorem. In either case, we have $\mu(\angle ACB) \le \mu(\angle ABC)$, and the proof is complete. ■

Theorem 44 (Triangle Inequality) *If A, B, and C are distinct points, then $AB + BC \ge AC$ and the equality holds if and only if $A - B - C$.*

Proof. See Exercise 1. ■

Theorem 45 (AAS) *If $\triangle ABC$ and $\triangle DEF$ are triangles such that $\angle ABC \cong \angle DEF$, $\angle BCA \cong \angle EFD$, and $\overline{AC} \cong \overline{DF}$, then $\triangle ABC \cong \triangle DEF$.*

Proof. See Exercise 2. ■

Definition 46 *A triangle is a **right triangle** if one of the interior angles is a right angle. The side opposite the right angle in a right triangle is the **hypotenuse** and the other two sides are the **legs**.*

Note that Side-Side-Angle (SSA) is not a valid triangle congruence theorem (see Exercise 3). However, when the corresponding congruent angles are right angles, SSA becomes a congruence theorem known as the *Hypotenuse-Leg Theorem* (HL). The reason for this name is that the right angle in a right triangle is always between its two legs. In the SSA situation for right triangles, however, the right angle (A) is not between the two sides (SS). Consequently, one side is the hypotenuse and the other is a leg.

Theorem 47 (Hypotenuse-Leg Theorem (HL)) *Let $\triangle ABC$ and $\triangle DEF$ be right triangles with right angles at the vertices C and F, respectively. If $\overline{AB} \cong \overline{DE}$ and $\overline{BC} \cong \overline{EF}$, then $\triangle ABC \cong \triangle DEF$.*

Proof. See Exercise 4. ■

The following theorem characterizes all of the points that lie on a perpendicular bisector. This theorem can be applied to prove our final triangle congruence theorem, namely, SSS.

Theorem 48 (Pointwise Characterization of Perpendicular Bisector) *Let A and B be distinct points. Then a point P lies on the perpendicular bisector of $\overline{AB}$ if and only if $PA = PB$.*

Proof. See Exercise 5. ■

Theorem 49 (SSS) *If $\triangle ABC$ and $\triangle DEF$ are triangles such that $\overline{AB} \cong \overline{DE}$, $\overline{BC} \cong \overline{EF}$, and $\overline{CA} \cong \overline{FD}$, then $\triangle ABC \cong \triangle DEF$.*

Proof. See Exercise 6. ■

We conclude this section with Theorem 51, which characterizes all of the points that lie on an angle bisector. But before we can state the theorem, we need the following definition:

Definition 50 *If l is a line and P is a point, the **distance** from P to l is the distance from P to the foot of the perpendicular from P to l.*

Theorem 51 (Pointwise Characterization of Angle Bisector) *Let A, B, and C be non-collinear points, and let P be a point in the interior of $\angle BAC$. Then P lies on the angle bisector of $\angle BAC$ if and only if the distances from P to lines $\overleftrightarrow{AB}$ and $\overleftrightarrow{AC}$ are equal.*

Proof. See Exercise 7. ■

Exercises

1. Prove Theorem 44 (Triangle Inequality).

2. Prove Theorem 45 (AAS).

3. Let $\triangle ABC$ and $\triangle DEF$ be triangles such that $\overline{AB} \cong \overline{DE}$, $\overline{BC} \cong \overline{EF}$, and $\angle ACB \cong \angle DFE$. Prove that $\angle BAC$ and $\angle EDF$ are congruent or supplementary.

4. Prove Theorem 47 (Hypotenuse-Leg Theorem (HL)).

5. Prove Theorem 48 (Pointwise Characterization of Perpendicular Bisector).

6. Prove Theorem 49 (SSS).

7. Prove Theorem 51 (Pointwise Characterization of Angle Bisector).

2.3 The Alternate Interior Angles Theorem and the Angle Sum Theorem

In this section, we discuss theorems involving parallel lines and related implications. We begin with the definitions of alternate interior and corresponding angles.

Definition 52 *Let l and l' be distinct lines, and let t be a transversal that cuts l at B and l' at B'. Let A and C be points on l such that $A - B - C$; let A' and C' be points on l' such that A' and A are on the same side of t and $A' - B' - C'$; let B'' be a point on t such that $B - B' - B''$ (see Figure 2.6).*

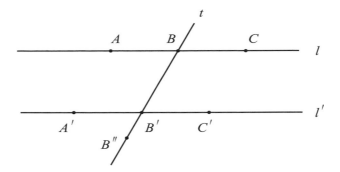

Figure 2.6. Definition 52.

1. *The angles* $\angle ABB'$ *and* $\angle C'B'B$ *are called* **alternate interior angles**. *Similarly, the angles* $\angle A'B'B$ *and* $\angle CBB'$ *are also alternate interior angles.*

2. *The angles* $\angle ABB'$ *and* $\angle A'B'B$ *are called* **non-alternate interior angles**. *Similarly, the angles* $\angle CBB'$ *and* $\angle C'B'B$ *are also non-alternate interior angles.*

3. *The angles* $\angle ABB'$ *and* $\angle A'B'B''$ *are called* **corresponding angles**. *There are four pairs of corresponding angles, each defined in the same manner.*

Theorem 53 (Alternate Interior Angles Theorem) *Let* l *and* l' *be distinct lines cut by a transversal* t. *Then* $l \parallel l'$ *if and only if alternate interior angles are congruent.*

Proof. Let l and l' be two distinct lines cut by a transversal t at point B on l and point B' on l'. Choose points A, A', C and C' as in Definition 52.

($\Leftarrow$) We prove the contrapositive. Assume that $l \nparallel l'$. Then by definition, l and l' intersect at some point D. Without loss of generality, assume that D and C are on the same side of t and consider $\triangle BB'D$ (see Figure 2.7). By the Exterior Angle Theorem, $\mu(\angle ABB') > \mu(\angle C'B'B)$ and the alternate interior angles $\angle ABB'$ and $\angle C'B'B$ are not congruent. By the same argument, the other pair of alternate interior angles $\angle A'B'B$ and $\angle CBB'$ are not congruent.

($\Rightarrow$) Assume that $l \parallel l'$. By the Angle Construction Postulate, there exists a line l'' through B' such that the alternate interior angles formed by l and l'' with transversal t are congruent (see Figure 2.7). Then $l \parallel l''$ by the converse proved above. Since B' is on both l' and l'', and both are parallel to l, $l' = l''$ by the Euclidean Parallel Postulate. Thus the alternate interior angles formed by l and l' with transversal t are congruent. ∎

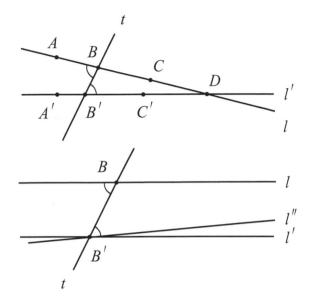

Figure 2.7. Proof of the Alternate Interior Angles Theorem.

One can apply the Vertical Angles Theorem to prove that the Alternate Interior Angles Theorem implies the Corresponding Angles Theorem:

Corollary 54 (Corresponding Angles Theorem) *Let l and l′ be distinct lines cut by a transversal t. Then l and l′ are parallel if and only if corresponding angles are congruent.*

Proof. See Exercise 1. ■

Similarly, one can apply the Supplement Postulate to prove that the Alternate Interior Angles Theorem implies the Non-alternate Interior Angles Theorem:

Corollary 55 (Non-alternate Interior Angles Theorem) *Let l and l′ be distinct lines cut by a transversal t. Then l and l′ are parallel if and only if non-alternate interior angles are supplementary.*

Proof. See Exercise 2. ■

Definition 56 *The **angle sum** of a triangle △ABC, denoted by σ(△ABC), is the sum of the measures of three interior angles.*

Another corollary of the Alternate Interior Angles Theorem is the following well-known theorem in Euclidean plane geometry:

Theorem 57 (Angle Sum Theorem) *For every triangle* $\triangle ABC$, *the angle sum* $\sigma(\triangle ABC) = 180°$.

Proof. Let $\triangle ABC$ be a triangle. By the Euclidean Parallel Postulate, there exists a line l through A such that $l \parallel \overleftrightarrow{BC}$. Choose points D and E on l such that $D - A - E$, $\angle DAB$ and $\angle ABC$ are alternate interior angles for $l \parallel \overleftrightarrow{BC}$ cut by transversal $\overleftrightarrow{AB}$, and $\angle EAC$ and $\angle ACB$ are alternate interior angles for $l \parallel \overleftrightarrow{BC}$ cut by transversal $\overleftrightarrow{AC}$ (see Figure 2.8). By the Alternate Interior Angles Theorem, $\angle DAB \cong \angle ABC$ so that $\mu(\angle DAB) = \mu(\angle ABC)$. Similarly, $\angle EAC \cong \angle ACB$ so that $\mu(\angle EAC) = \mu(\angle ACB)$.

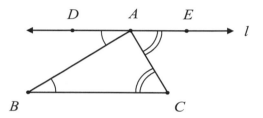

Figure 2.8. Proof of the Angle Sum Theorem.

By the Angle Addition Postulate, $\mu(\angle BAC) + \mu(\angle DAB) = \mu(\angle DAC)$. By the Supplement Postulate, $\mu(\angle DAC) + \mu(\angle EAC) = 180°$. Then by substitution,

$$\mu(\angle BAC) + \mu(\angle DAB) + \mu(\angle EAC) = 180°.$$

Again by substitution,

$$\mu(\angle BAC) + \mu(\angle ABC) + \mu(\angle ACB) = 180°.$$

Therefore the angle sum $\sigma(\triangle ABC) = 180°$ and the proof is complete. ∎

We conclude this section with two more theorems. The first lists some properties of parallelograms, and the second establishes the transitivity of parallelism. We leave the proofs of both theorems as exercises for the reader.

Definition 58 *Let* A, B, C, *and* D *be points no three of which are collinear.*

1. *The* **quadrilateral** $\square ABCD$ *is the union of the four segments* $\overline{AB}, \overline{BC}$, $\overline{CD}$, *and* $\overline{DA}$. *The points* A, B, C, *and* D *are the* **vertices** *of the quadrilateral, the segments* $\overline{AB}, \overline{BC}, \overline{CD}$, *and* $\overline{DA}$ *are the* **sides** *of the quadrilateral, and the segments* $\overline{AC}$ *and* $\overline{BD}$ *are the* **diagonals** *of the quadrilateral.*

2. *A quadrilateral* $\square ABCD$ *is* **simple** *if it is not self-intersecting, i.e., two sides have either no point in common or exactly one endpoint in common; otherwise* $\square ABCD$ *is a* **crossed** *quadrilateral (see Figure 2.9).*

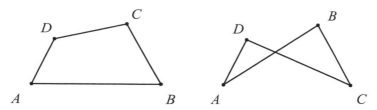

Figure 2.9. Simple (left) and crossed (right) quadrilaterals □*ABCD*.

Definition 59 *A quadrilateral* □*ABCD* *is a* ***parallelogram*** *if* $\overleftrightarrow{AB}$ ∥ $\overleftrightarrow{CD}$ *and* $\overleftrightarrow{AD}$ ∥ $\overleftrightarrow{BC}$. *A parallelogram with four congruent sides is a* ***rhombus***.

Theorem 60 (Properties of Parallelograms) *Let* □*ABCD* *be a parallelogram. Then*

a. △*ABC* ≅ △*CDA and* △*ABD* ≅ △*CDB*,

b. *opposite sides are congruent, i.e.,* $\overline{AB}$ ≅ $\overline{CD}$ *and* $\overline{BC}$ ≅ $\overline{DA}$,

c. *opposite angles are congruent, i.e.,* ∠*DAB* ≅ ∠*BCD and* ∠*ABC* ≅ ∠*CDA, and*

d. *the diagonals* $\overline{AC}$ *and* $\overline{BD}$ *bisect each other.*

Theorem 61 (Transitivity of Parallelism) *Let l, m, and n be lines. If l* ∥ *m and m* ∥ *n, then l* ∥ *n.*

Exercises

1. Prove Corollary 54 (Corresponding Angles Theorem).

2. Prove Corollary 55 (Non-alternate Interior Angles Theorem).

3. Prove that the measure of an exterior angle for a triangle is equal to the sum of the measures of two non-adjacent interior angles.

4. Prove Theorem 60 (Properties of Parallelograms).

5. Prove Theorem 61 (Transitivity of Parallelism).

2.4   Similar Triangles

Similar triangles will play a fundamental role in our study of similarities in transformational plane geometry. By definition, two triangles are similar if all three pairs of corresponding angles are congruent. In this section, we develop some alternative characterizations of triangle similarity.

Definition 62 *Triangles $\triangle ABC$ and $\triangle DEF$ are* **similar,** *written $\triangle ABC \sim \triangle DEF$, if $\angle ABC \cong \angle DEF$, $\angle BCA \cong \angle EFD$, and $\angle CAB \cong \angle FDE$.*

Theorem 63 (AA) *Two triangles are similar if two pairs of corresponding angles are congruent.*

Proof. See Exercise 1. ∎

The next theorem, which is a standard result in Euclidean plane geometry, will be applied in the proof of Theorem 65, the Similar Triangles Theorem.

Theorem 64 *Let $\triangle ABC$ be a triangle, let B' be a point on $\overline{AB}$, and let C' be a point on $\overline{AC}$. Then*

$$\frac{AB}{AB'} = \frac{AC}{AC'} \text{ if and only if } \overleftrightarrow{BC} \parallel \overleftrightarrow{B'C'}.$$

Proof. Let $\triangle ABC$ be a triangle, let B' be a point on $\overline{AB}$, and let C' be a point on $\overline{AC}$.

($\Leftarrow$) First assume that $\overleftrightarrow{BC} \parallel \overleftrightarrow{B'C'}$. We consider three cases:

Case 1: $\frac{AB}{AB'} = 2$. Then $AB' = B'B$. By the Euclidean Parallel Postulate, there is a unique line m through point C' and parallel to $\overleftrightarrow{AB}$. By Pasch's Axiom, m cuts $\overline{BC}$ at some point D (see Figure 2.10). By definition, $\square BB'C'D$ is a parallelogram. By Theorem 60 part 2, $B'B = C'D$. Hence $AB' = C'D$. By the Corresponding Angles Theorem, $\angle B'AC' \cong \angle DC'C$ and $\angle AC'B' \cong \angle C'CD$. Thus $\triangle AB'C' \cong \triangle C'DC$ by AAS and $AC' = C'C$ by CPCTC. Therefore $\frac{AC}{AC'} = 2$, and it follows that $\frac{AB}{AB'} = \frac{AC}{AC'}$.

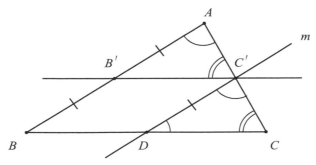

Figure 2.10. Proof of Theorem 64, Case 1.

<u>Case 2</u>: $\frac{AB}{AB'} = \frac{p}{q}$ for some $p, q \in \mathbb{N}$ with $p > q$. By the Ruler Postulate, there exist points $B_1, B_2, \ldots, B_p$ on $\overline{AB}$ such that $B_i B_{i+1} = AB_1 = \frac{AB}{p}$ for each $i = 1, 2, \ldots, p - 1$. Note that $B_q = B'$ and $B_p = B$. By the Euclidean Parallel Postulate, for each $i = 1, 2, \ldots, p$, there exists a unique line l_i through point B_i and parallel to $\overleftrightarrow{BC}$ (see Figure 2.11). By Pasch's Axiom, l_i cuts $\overline{AC}$ at some point C_i, and it follows that $C_q = C'$ and $C_p = C$. Now for each $i = 1, 2, \ldots, p - 1$, let m_i be the unique line through point C_i and parallel to $\overleftrightarrow{AB}$ given by the Euclidean Parallel Postulate. By Pasch's Axiom, m_i cuts $\overline{B_{i+1}C_{i+1}}$ at some point D_{i+1}. By the argument in Case 1, $\triangle AB_1 C_1 \cong \triangle C_i D_{i+1} C_{i+1}$ and consequently $C_i C_{i+1} = AC_1 = \frac{AC}{p}$ for each $i = 1, 2, \ldots, p - 1$. Therefore $AC = pAC_1$ and $AC' = qAC_1$ so that

$$\frac{AB}{AB'} = \frac{p}{q} = \frac{pAC_1}{qAC_1} = \frac{AC}{AC'}.$$

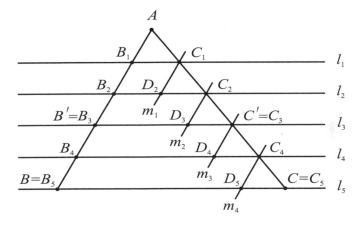

Figure 2.11. Proof of Theorem 64, Case 2 (for $p = 5, q = 3$).

<u>Case 3</u>: $\frac{AB}{AB'}$ and $\frac{AC}{AC'}$ are real numbers. Since $A - B' - B$ and $A - C' - C$, $\frac{AB}{AB'} > 1$ and $\frac{AC}{AC'} > 1$. Let d be any rational number greater than 1 and assume that $\frac{AB}{AB'} > d$. We claim that $\frac{AC}{AC'} > d$. By the Ruler Postulate, there is a point D on $\overrightarrow{AB}$ such that $\frac{AD}{AB'} = d$. Thus $B' - D - B$. By the Euclidean Parallel Postulate, there is a unique line n through D and parallel to $\overleftrightarrow{BC}$ (see Figure 2.12). By Pasch's Axiom, n cuts $\overline{AC}$ at some point E. By Case 2, $\frac{AE}{AC'} = \frac{AD}{AB'} = d$. Since $AC > AE$, we have $\frac{AC}{AC'} > \frac{AE}{AC'}$ so that $\frac{AC}{AC'} > d$ as claimed. A similar argument shows that $\frac{AC}{AC'} > d$ implies $\frac{AB}{AB'} > d$. Thus for every rational number $d > 1$, we have $\frac{AB}{AB'} > d$ if and only if $\frac{AC}{AC'} > d$. Consequently, $\frac{AB}{AB'} = \frac{AC}{AC'}$ by properties of the real numbers.

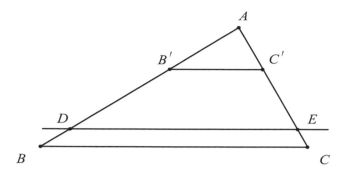

Figure 2.12. Proof of Theorem 64, Case 3.

($\Rightarrow$) Assume that $\frac{AB}{AB'} = \frac{AC}{AC'}$. By the Euclidean Parallel Postulate, there is a line t through B' and parallel to $\overleftrightarrow{BC}$. By Pasch's Axiom, t cuts $\overline{AC}$ at some point C''. By the previous argument, $\frac{AB}{AB'} = \frac{AC}{AC''}$, and by substitution, $\frac{AC}{AC'} = \frac{AC}{AC''}$. Thus $AC' = AC''$ and $C' = C''$. Consequently $\overleftrightarrow{B'C'} = t$ and $\overleftrightarrow{BC} \parallel \overleftrightarrow{B'C'}$. ∎

We are ready to prove the Similar Triangles Theorem, which asserts that two triangles are similar if and only if the ratios of the lengths of their corresponding sides are equal. In fact, some authors use this characterization to define similar triangles.

Theorem 65 (Similar Triangles Theorem) *Two triangles are similar if and only if the ratios of the lengths of their corresponding sides are equal, i.e.,*

$$\triangle ABC \sim \triangle DEF \ \text{if and only if} \ \frac{AB}{DE} = \frac{AC}{DF} = \frac{BC}{EF}.$$

Proof. Given a pair of triangles $\triangle ABC$ and $\triangle DEF$, we consider two cases.

Case 1: $AB = DE$. ($\Rightarrow$) Assume that $\triangle ABC \sim \triangle DEF$. By definition, $\angle CAB \cong \angle FDE$ and $\angle ABC \cong \angle DEF$. Then by ASA, $\triangle ABC \cong \triangle DEF$. Hence $AC = DF$ and $BC = EF$ by CPCTC. Thus, $\frac{AB}{DE} = \frac{AC}{DF} = \frac{BC}{EF}$.

($\Leftarrow$) Assume $\frac{AB}{DE} = \frac{AC}{DF} = \frac{BC}{EF}$. Since $AB = DE$, we have $AC = DF$ and $BC = EF$ by algebra, so that $\triangle ABC \cong \triangle DEF$ by SSS. Therefore $\triangle ABC \sim \triangle DEF$.

Case 2: $AB \neq DE$. Without loss of generality, assume $AB > DE$. Before constructing the proof, we make some preliminary observations. By the Ruler Postulate, there exists a point B' on $\overline{AB}$ such that $AB' = DE$. By the Euclidean Parallel Postulate, there exists a line l through B' and parallel to $\overleftrightarrow{BC}$. By Pasch's Axiom, l cuts $\overline{AC}$ at some point C'. By the Corresponding Angles Theorem, $\angle ABC \cong \angle AB'C'$ and $\angle ACB \cong \angle AC'B'$ (see Figure 2.13).

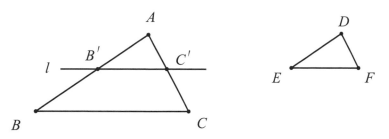

Figure 2.13. Proof of Theorem 65, Case 2.

($\Rightarrow$) Assume $\triangle ABC \sim \triangle DEF$. By definition, $\angle ABC \cong \angle DEF$ and $\angle ACB \cong \angle DFE$. Hence $\angle AB'C' \cong \angle DEF$ and $\angle AC'B' \cong \angle DFE$ so that $\triangle AB'C' \cong \triangle DEF$ by AAS. Thus $AC' = DF$ by CPCTC. Since $\overleftrightarrow{BC} \parallel \overleftrightarrow{B'C'}$, $\frac{AB}{AB'} = \frac{AC}{AC'}$ by Theorem 64. By substitution, $\frac{AB}{DE} = \frac{AC}{DF}$. The proof of $\frac{AC}{DF} = \frac{BC}{EF}$ is identical by switching roles of the vertices of the similar triangles. Thus $\frac{AB}{DE} = \frac{AC}{DF} = \frac{BC}{EF}$.

($\Leftarrow$) Assume $\frac{AB}{DE} = \frac{AC}{DF} = \frac{BC}{EF}$. Since $\angle ABC \cong \angle AB'C'$ and $\angle ACB \cong \angle AC'B'$, $\triangle ABC \sim \triangle AB'C'$ by AA. By the above argument, $\frac{AB}{AB'} = \frac{AC}{AC'} = \frac{BC}{B'C'}$. This together with $\frac{AB}{DE} = \frac{AC}{DF} = \frac{BC}{EF}$ and $AB' = DE$, we conclude that $AC' = DE$ and $B'C' = EF$. Thus $\triangle AB'C' \cong \triangle DEF$ by SSS, and $\triangle ABC \sim \triangle DEF$ by substitution. ∎

We conclude this section with the Side-Angle-Side criterion for similar triangles. This result specializes to the Side-Angle-Side Postulate for congruent triangles when $\frac{AB}{AC} = \frac{DE}{DF} = 1$.

Theorem 66 (SAS for Similar Triangles) *If $\triangle ABC$ and $\triangle DEF$ are triangles such that $\angle CAB \cong \angle FDE$ and $\frac{AB}{AC} = \frac{DE}{DF}$, then $\triangle ABC \sim \triangle DEF$.*

Proof. See Exercise 2. ∎

Exercises

1. Prove Theorem 63 (AA).

2. Prove Theorem 66 (SAS for Similar Triangles).

Chapter 3

Introduction to Transformations, Isometries, and Similarities

Any problem in geometry can easily be reduced to such terms that a knowledge of the lengths of certain straight lines is sufficient for its construction.

René Descartes (1596–1650)

In grades 9–12 all students should apply transformations and use symmetry to analyze mathematical situations.

Principles and Standards for School Mathematics
National Council of Teachers of Mathematics (2000)

This chapter introduces some of the key players in our story – the length-preserving and length-ratio-preserving transformations of the Euclidean plane $\mathbb{R}^2$.

3.1 Transformations

We begin with the definition of a transformation of $\mathbb{R}^2$ and review some of its basic properties. Functions from $\mathbb{R}^2$ to $\mathbb{R}^2$ are denoted by lower case Greek letters such as $\alpha, \beta, \gamma, \ldots$.

Definition 67 *A **transformation** of the plane is a function* $\alpha : \mathbb{R}^2 \to \mathbb{R}^2$ *with domain* $\mathbb{R}^2$.

Example 68 The **identity transformation** $\iota : \mathbb{R}^2 \to \mathbb{R}^2$ is defined by $\iota(P) = P$ for every point P.

Recall that we defined injective, surjective, and bijective functions in Definition 4. For the sake of clarity, we restate these definitions for transformations of the plane.

Definition 69 *A transformation* $\alpha : \mathbb{R}^2 \to \mathbb{R}^2$ *is **injective** (or **one-to-one**) if distinct points have distinct images, i.e., if $P \neq Q$ then $\alpha(P) \neq \alpha(Q)$.*

Example 70 The transformation $\beta\left(\begin{bmatrix} x \\ y \end{bmatrix}\right) = \begin{bmatrix} x^2 \\ y \end{bmatrix}$ fails to be injective because $\beta\left(\begin{bmatrix} -1 \\ 1 \end{bmatrix}\right) = \begin{bmatrix} 1 \\ 1 \end{bmatrix} = \beta\left(\begin{bmatrix} 1 \\ 1 \end{bmatrix}\right).$

Our next example illustrates how to verify injectivity using the contrapositive condition: *if $\alpha(P) = \alpha(Q)$, then $P = Q$.*

Example 71 To show that the transformation $\gamma\left(\begin{bmatrix} x \\ y \end{bmatrix}\right) = \begin{bmatrix} x+2y \\ 2x-y \end{bmatrix}$ is injective, assume that $\gamma\left(\begin{bmatrix} a \\ b \end{bmatrix}\right) = \gamma\left(\begin{bmatrix} c \\ d \end{bmatrix}\right).$ Then

$$\begin{bmatrix} a + 2b \\ 2a - b \end{bmatrix} = \begin{bmatrix} c + 2d \\ 2c - d \end{bmatrix},$$

and by equating x and y coordinates we obtain

$$a + 2b = c + 2d$$
$$2a - b = 2c - d.$$

A simple calculation now shows that $b = d$ and $a = c$ so that $\begin{bmatrix} a \\ b \end{bmatrix} = \begin{bmatrix} c \\ d \end{bmatrix}$.

Definition 72 *A transformation* $\alpha : \mathbb{R}^2 \to \mathbb{R}^2$ *is **surjective** (or **onto**) if the range of α is $\mathbb{R}^2$, i.e., given any point $Q \in \mathbb{R}^2$, there is some point $P \in \mathbb{R}^2$ such that $\alpha(P) = Q$.*

Example 73 The transformation $\beta\left(\begin{bmatrix} x \\ y \end{bmatrix}\right) = \begin{bmatrix} x^2 \\ y \end{bmatrix}$ discussed in Example 70 fails to be surjective because there is no point $\begin{bmatrix} x \\ y \end{bmatrix} \in \mathbb{R}^2$ such that $\beta\left(\begin{bmatrix} x \\ y \end{bmatrix}\right) = \begin{bmatrix} -1 \\ 0 \end{bmatrix}$, for example.

Example 74 To determine whether or not the injective transformation $\gamma\left(\begin{bmatrix} x \\ y \end{bmatrix}\right) = \begin{bmatrix} x+2y \\ 2x-y \end{bmatrix}$ discussed in Example 71 is also surjective, we must answer the following question: Given any point $Q = \begin{bmatrix} c \\ d \end{bmatrix}$, do there exist choices for x and y such that $\gamma\left(\begin{bmatrix} x \\ y \end{bmatrix}\right) = \begin{bmatrix} x+2y \\ 2x-y \end{bmatrix} = \begin{bmatrix} c \\ d \end{bmatrix}$? The answer is yes if the following linear system of equations has a solution:

$$x + 2y = c$$
$$2x - y = d,$$

which indeed it does since the determinant $\begin{vmatrix} 1 & 2 \\ 2 & -1 \end{vmatrix} = -1 - 4 = -5 \neq 0$. Solving for x and y in terms of c and d we find that

$$x = \tfrac{1}{5}c + \tfrac{2}{5}d$$
$$y = \tfrac{2}{5}c - \tfrac{1}{5}d.$$

Therefore $\gamma\left(\begin{bmatrix} \frac{1}{5}c + \frac{2}{5}d \\ \frac{2}{5}c - \frac{1}{5}d \end{bmatrix}\right) = \begin{bmatrix} c \\ d \end{bmatrix}$ and γ is surjective.

Definition 75 *A **bijective** transformation is both injective and surjective.*

Example 76 The transformation γ discussed in Examples 71 and 74 is bijective.

Definition 77 *Let α be a bijective transformation. The **inverse of** α is the function α^{-1} defined in the following way: if P is any point and Q is the unique point such that $\alpha(Q) = P$, then $\alpha^{-1}(P) = Q$.*

In Exercise 5 you will be asked to prove that the inverse of a bijective transformation is a bijective transformation. Using this fact it is easy to prove:

Proposition 78 *Let α and β be bijective transformations. Then $\beta = \alpha^{-1}$ if and only if $\alpha \circ \beta = \beta \circ \alpha = \iota$.*

Proof. ($\Leftarrow$) Suppose $\beta = \alpha^{-1}$. Given any point Q, let $P = \alpha(Q)$. Then $(\beta \circ \alpha)(Q) = \beta(\alpha(Q)) = \beta(P) = \alpha^{-1}(P) = Q = \iota(Q)$ so that $\beta \circ \alpha = \iota$. Similarly, given any point P, let $Q = \alpha^{-1}(P)$. Then $(\alpha \circ \beta)(P) = \alpha(\alpha^{-1}(P)) = \alpha(Q) = P = \iota(P)$ so that $\alpha \circ \beta = \iota$.

($\Rightarrow$) Suppose $\alpha \circ \beta = \beta \circ \alpha = \iota$. Given any point P, let $Q = \alpha^{-1}(P)$. Then $\alpha^{-1}(P) = Q = \iota(Q) = (\beta \circ \alpha)(Q) = \beta(\alpha(Q)) = \beta(P)$. Therefore $\beta = \alpha^{-1}$. ∎

Exercises

1. Which of the following transformations are injective? Which are surjective? Which are bijective?

$$\alpha\left(\begin{bmatrix} x \\ y \end{bmatrix}\right) = \begin{bmatrix} x^3 \\ y \end{bmatrix} \quad \beta\left(\begin{bmatrix} x \\ y \end{bmatrix}\right) = \begin{bmatrix} \cos x \\ \sin y \end{bmatrix} \quad \gamma\left(\begin{bmatrix} x \\ y \end{bmatrix}\right) = \begin{bmatrix} x^3 - x \\ y \end{bmatrix}$$

$$\delta\left(\begin{bmatrix} x \\ y \end{bmatrix}\right) = \begin{bmatrix} 2x \\ 3y \end{bmatrix} \quad \varepsilon\left(\begin{bmatrix} x \\ y \end{bmatrix}\right) = \begin{bmatrix} -x \\ x+3 \end{bmatrix} \quad \eta\left(\begin{bmatrix} x \\ y \end{bmatrix}\right) = \begin{bmatrix} 3y \\ x+2 \end{bmatrix}$$

$$\rho\left(\begin{bmatrix} x \\ y \end{bmatrix}\right) = \begin{bmatrix} \sqrt[3]{x} \\ e^y \end{bmatrix} \quad \sigma\left(\begin{bmatrix} x \\ y \end{bmatrix}\right) = \begin{bmatrix} -x \\ -y \end{bmatrix} \quad \tau\left(\begin{bmatrix} x \\ y \end{bmatrix}\right) = \begin{bmatrix} x+2 \\ y-3 \end{bmatrix}$$

2. Prove that the composition of bijective transformations is a bijective transformation. Thus the composition of bijective transformations is *closed* with respect to function composition.

3. Prove that if α is a transformation, then $\alpha \circ \iota = \iota \circ \alpha = \alpha$. Thus the identity transformation ι is an *identity element* for the set of all transformations with respect to functions composition.

4. Prove that function composition is associative, i.e., if f, g, and h are functions, then $f \circ (g \circ h) = (f \circ g) \circ h$. In particular, the composition of bijective transformations is *associative*.

5. Prove that the inverse of a bijective transformation is a bijective transformation. Thus the set of bijective transformations contains the *inverses* of its elements. (Remark: This exercise, together with Exercises 3.1.2, 3.1.3, and 3.1.4, establishes the fact that the set of all bijective transformations is a *group* under function composition.)

6. Let α and β be bijective transformations. Prove that $(\alpha \circ \beta)^{-1} = \beta^{-1} \circ \alpha^{-1}$, i.e., the inverse of a composition of bijective transformations is the composition of their inverses in reverse order.

3.2   Isometries and Similarities

One of the primary goals of this exposition is to find and classify the isometries and similarities of the plane. In this section we define these transformations and establish some of their basic properties.

Let P and Q be points. Recall that the symbol PQ denotes the distance from P to Q. From now on, we use PQ specifically to denote the Euclidean distance. To express PQ in terms of coordinates, let $P = \begin{bmatrix} x_1 \\ y_1 \end{bmatrix}$ and $Q = \begin{bmatrix} x_2 \\ y_2 \end{bmatrix}$. Then

$$PQ = \sqrt{(x_2 - x_1)^2 + (y_2 - y_1)^2}.$$

Definition 79 *Let $\alpha : \mathbb{R}^2 \to \mathbb{R}^2$ be a transformation, let P and Q be points, and let $P' = \alpha(P)$ and $Q' = \alpha(Q)$. Then α is an **isometry** if $P'Q' = PQ$. Thus an isometry is a distance-preserving transformation of the plane.*

Example 80 The identity transformation ι is an isometry since $P' = \iota(P) = P$ and $Q' = \iota(Q) = Q$ for all P and Q so that $P'Q' = PQ$. On the other hand, the transformation $\gamma\left(\begin{bmatrix} x \\ y \end{bmatrix}\right) = \begin{bmatrix} x+2y \\ 2x-y \end{bmatrix}$ discussed in Examples 71 and 74 is not an isometry. To see this, consider the points $O = \begin{bmatrix} 0 \\ 0 \end{bmatrix}$ and $E = \begin{bmatrix} 1 \\ 0 \end{bmatrix}$ and their images $O' = \gamma(O) = \begin{bmatrix} 0 \\ 0 \end{bmatrix}$ and $E' = \gamma(E) = \begin{bmatrix} 1 \\ 2 \end{bmatrix}$. Then $O'E' = \sqrt{5} \neq 1 = OE$.

In Exercise 1 we ask the reader to supply the proof of the following fact:

Proposition 81 *Isometries are injective transformations.*

Isometries are also surjective; however, the proof is somewhat technical and appears in the appendix at the end of this chapter.

In Definition 18 of Chapter 1 we defined an angle to be the union of two rays $\overrightarrow{AB}$ and $\overrightarrow{AC}$ such that A is not between B and C. Note that in that definition the symbols $\angle BAC$ and $\angle CAB$ represent the same angle. These two symbols, however, are distinguished in the following definition. Specifically, the symbol $\angle BAC$ denotes the angle directed from $\overrightarrow{AB}$ and $\overrightarrow{AC}$, while $\angle CAB$ denotes the angle directed from $\overrightarrow{AC}$ and $\overrightarrow{AB}$. Furthermore, we also consider $\angle BAC$ to be an angle when A is between B and C.

Definition 82 *Let A, B, and C be points such that $B \neq A$ and $C \neq A$, and choose a unit of measure. If A, B, and C are non-collinear, let $s \in (0, \pi)$ be the length of the unit arc subtended by $\angle BAC$ and directed from $\overrightarrow{AB}$ to $\overrightarrow{AC}$. The **measure of the directed angle** $\angle BAC$ is the real number*

$$
\theta = \begin{cases}
0, & \text{if } \overrightarrow{AB} = \overrightarrow{AC} \\
\pi, & \text{if } B - A - C \\
s, & \text{if the unit subtended arc is directed counterclockwise} \\
-s, & \text{if the unit subtended arc is directed clockwise.}
\end{cases}
$$

*The **degree measure of the directed angle** $\angle BAC$, denoted by $m\angle BAC$, is the real number $\Theta = \frac{180}{\pi}\theta \in (-180, 180]$.*

Degree measure will be used exclusively throughout this text. Note that $m\angle BAC = -m\angle CAB$ whenever A, B, and C are non-collinear. We also want to emphasize that $\mu(\angle BAC) = |m\angle BAC|$ satisfies all of the conditions in the Protractor Postulate, and $\angle ABC \cong \angle DEF$ if and only if $|m\angle ABC| = |m\angle DEF|$.

Definition 83 *Two real numbers Θ and Φ are **congruent modulo** 360 if $\Theta - \Phi = 360k$ for some $k \in \mathbb{Z}$, in which case we write $\Theta \equiv \Phi$. The symbol $\Theta°$ denotes the congruence class of Θ, i.e.,*

$$
\Theta° = \{\Phi \in \mathbb{R} \mid \Theta - \Phi = 360k, \text{ for some } k \in \mathbb{Z}\}.
$$

Note that the congruence class $\Theta°$ contains exactly one real number in the interval $(-180, 180]$. For example, $370° \cap (-180, 180] = 10$. When $\Theta \equiv \Phi$, the congruence classes $\Theta° = \Phi°$ are equal sets. Thus $370° = 10°$.

Definition 84 *Let $m\angle ABC = \Theta$ and $m\angle DEF = \Phi$. Define the **angle sum** $\Theta + \Phi$ to be the unique real number $\Psi \in (-180, 180]$ such that $\Psi \equiv \Theta + \Phi$. Define $-\Theta° := (-\Theta)°$ and $\Theta° + \Phi° := (\Theta + \Phi)°$.*

For example, $-10° = (-10)°$ and $370° + 10° = 20°$. If $m\angle ABC = m\angle CBD = 120$, then $m\angle ABC + m\angle CBD = 240$, while $m\angle ABD = -120$. This is consistent with the fact that the congruence class $2(120)° = (2(120))° = 240°$ is represented by $-120 \in (-180, 180]$.

Since distinct elements of $(-180, 180]$ are not congruent $(\bmod\, 360)$ we have $m\angle ABC \equiv m\angle A'B'C'$ if and only if $m\angle ABC = m\angle A'B'C'$. However, $m\angle ABC \equiv m\angle A'B'C'$ implies $\angle ABC \cong \angle A'B'C'$, but not conversely.

Definition 85 *Let P and Q be distinct points. The **circle centered at P containing Q** is the set of points*

$$
P_Q := \{X \mid PX = PQ\}.
$$

*The **radius** of P_Q is the distance PQ. If R is a point on P_Q such that $R - P - Q$, then $\overline{RQ}$ is called a **diameter** of P_Q.*

If D is any point on A_B distinct from B and C, the following proposition asserts that $\angle BDC$ is a right angle. Thus an inscribed triangle is a right triangle whenever one of its sides is a diameter.

Proposition 86 *If B, C and D are distinct points on a circle centered at A, then $m\angle BAC \equiv 2m\angle BDC$.*

Proof. When $\square ABDC$ is a simple quadrilateral, the points B and C lie on opposite sides of $\overleftrightarrow{AD}$. Label the interior angles of $\triangle ABD$ and $\triangle ACD$ as in Figure 3.1: $\angle 1 := \angle ABD$, $\angle 2 := \angle BDA$, $\angle 3 := \angle ADC$, $\angle 4 := \angle DCA$, $\angle 5 := \angle DAB$, and $\angle 6 := \angle CAD$. Since the measures of these interior angles have the same sign, $m\angle 1 = m\angle 2$, $m\angle 3 = m\angle 4$, and

$$m\angle BDC = m\angle 2 + m\angle 3 = m\angle 1 + m\angle 4.$$

Furthermore, the interior angle sum of $\square ABDC$ is $360 \equiv m\angle 1 + m\angle 2 + m\angle 3 + m\angle 4 + m\angle 5 + m\angle 6 = 2m\angle 2 + 2m\angle 3 + m\angle 5 + m\angle 6$, and it follows that

$$m\angle BAC \equiv 360 - m\angle 5 - m\angle 6 \equiv 2\,(m\angle 2 + m\angle 3) = 2m\angle BDC.$$

The crossed quadrilateral case is left as an exercise for the reader. ∎

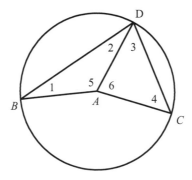

Figure 3.1. Proof of Proposition 86 when $\square ABDC$ is simple.

To see that Proposition 86 only holds $\bmod\,360$, consider $\square ABDC$ with $m\angle 2 = m\angle 3 = 60$. Then $m\angle BAC = -120 \equiv 240 = 2m\angle BDC$.

Definition 87 *Let P be a point. Two or more lines or circles are **concurrent** at P if each line or circle passes through P.*

Here is another useful fact, which applies in our proof that isometries are surjective:

Proposition 88 *Three concurrent circles with non-collinear centers have a unique point of concurrency.*

Proof. We prove the contrapositive. Suppose that three circles centered at points A, B, and C are concurrent at distinct points P and Q. Then $AP = AQ$. By Theorem 48, point A lies on the perpendicular bisector of $\overline{PQ}$. Similarly, B and C also lie on the perpendicular bisector of $\overline{PQ}$. Thus A, B, and C are collinear. ∎

Proposition 89 *Let α and β be isometries.*

1. *The composition $\alpha \circ \beta$ is an isometry.*

2. *$\alpha \circ \iota = \iota \circ \alpha = \alpha$, i.e., the identity transformation acts as an identity element.*

3. *α^{-1} is an isometry, i.e., the inverse of an isometry is an isometry.*

Proof. The proofs are left as exercises for the reader. ∎

Definition 90 *A bijective transformation $\phi : \mathbb{R}^2 \to \mathbb{R}^2$ is a **collineation** if it sends lines to lines.*

Suppose a given transformation α is known to be a collineation. Given the equation of a line l, we wish to compute the equation of its image line l'. One successful strategy is to arbitrarily choose two distinct points on l then use the equations of α to compute the coordinates of their image points. Since these image points are on l', we can find the equation of l' using the two-point formula for the equation of a line. On the other hand, when it is easy to solve the equations of α for x and y in terms of x' and y', a more efficient strategy is to assume $\begin{bmatrix} x' \\ y' \end{bmatrix} \in l'$, then solve for x and y, and substitute in the equation of l. This procedure immediately produces the equation of l'. Our next example demonstrates both methods.

Example 91 The transformation $\alpha\left(\begin{bmatrix} x \\ y \end{bmatrix}\right) = \begin{bmatrix} y+1 \\ 2-x \end{bmatrix}$ is known to be a collineation. Let l be the line with equation $X + 2Y - 6 = 0$. Choose two arbitrary points on l, say $\begin{bmatrix} 6 \\ 0 \end{bmatrix}$ and $\begin{bmatrix} 0 \\ 3 \end{bmatrix}$. Using the equations of α we find that $\alpha\left(\begin{bmatrix} 6 \\ 0 \end{bmatrix}\right) = \begin{bmatrix} 1 \\ -4 \end{bmatrix}$ and $\alpha\left(\begin{bmatrix} 0 \\ 3 \end{bmatrix}\right) = \begin{bmatrix} 4 \\ 2 \end{bmatrix}$. Thus the equation of l' is $2X - Y - 6 = 0$. On the other hand, we have

$$x' = y + 1 \Rightarrow y = x' - 1 \text{ and } y' = 2 - x \Rightarrow x = 2 - y'.$$

Since $\begin{bmatrix} x \\ y \end{bmatrix} \in l$, we have $x + 2y - 6 = 0$. Substituting $2 - y'$ for x and $x' - 1$ for y gives $(2 - y') + 2(x' - 1) - 6 = 0$, which simplifies to $2x' - y' - 6 = 0$. So the equation of l' is $2X - Y - 6 = 0$, which agrees with our previous calculation.

Proposition 92 *Let α be an isometry. Given distinct points A, B, and C, let $A' = \alpha(A)$, $B' = \alpha(B)$, and $C' = \alpha(C)$. Then α*

1. *is a collineation.*

2. *preserves betweenness, i.e., if $A - B - C$, then $A' - B' - C'$.*

3. *preserves angle measure up to sign, i.e., $\angle A'B'C' \cong \angle ABC$.*

4. *sends circles to circles, i.e., $\alpha(A_B) = A'_{B'}$.*

Proof. The proofs are left as exercises for the reader. ∎

Definition 93 *Let $r > 0$. A **similarity of ratio** r is a transformation $\alpha :$ $\mathbb{R}^2 \to \mathbb{R}^2$ that preserves the image-to-preimage distance ratio r, i.e., for all points P and Q, if $P' = \alpha(P)$ and $Q' = \alpha(Q)$, then $P'Q' = rPQ$ (see Figure 3.2).*

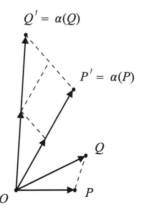

Figure 3.2. A similarity α of ratio 2.

Example 94 If α is an isometry, P and Q are points, $P' = \alpha(P)$ and $Q' = \alpha(Q)$, then $P'Q' = PQ$. Hence an isometry is a similarity of ratio $r = 1$. It is easy to show that for each $r > 0$, the transformation defined by $\beta\left(\begin{bmatrix} x \\ y \end{bmatrix}\right) = \begin{bmatrix} rx \\ ry \end{bmatrix}$ is a similarity of ratio r (see Exercise 12).

Exercise 14 asks the reader to supply the proof of the following fact:

Proposition 95 *Similarities are bijective transformations.*

Exercises

1. Prove that isometries are injective transformations.

2. Prove Proposition 89.

3. Let l be the line with equation $2X + 3Y + 4 = 0$. Using the fact that each of the following transformations is a collineation, find the equation of the image line l'.

(a) $\alpha\left(\begin{bmatrix} x \\ y \end{bmatrix}\right) = \begin{bmatrix} -x \\ y \end{bmatrix}$.

(b) $\beta\left(\begin{bmatrix} x \\ y \end{bmatrix}\right) = \begin{bmatrix} x \\ -y \end{bmatrix}$.

(c) $\gamma\left(\begin{bmatrix} x \\ y \end{bmatrix}\right) = \begin{bmatrix} -x \\ -y \end{bmatrix}$.

(d) $\delta\left(\begin{bmatrix} x \\ y \end{bmatrix}\right) = \begin{bmatrix} 5-x \\ 10-y \end{bmatrix}$.

(e) $\varepsilon\left(\begin{bmatrix} x \\ y \end{bmatrix}\right) = \begin{bmatrix} x+1 \\ y-3 \end{bmatrix}$.

4. Find an example of a bijective transformation that is not a collineation.

5. Which of the transformations in Exercise 3.1.1 are collineations? For each transformation that is a collineation, find the equation of the image of the line l with equation $aX + bY + c = 0$.

6. Consider the collineation $\alpha\left(\begin{bmatrix} x \\ y \end{bmatrix}\right) = \begin{bmatrix} 3y \\ x-y \end{bmatrix}$ and the line l' whose equation is $3X - Y + 2 = 0$. Find the equation of the line l such that $\alpha(l) = l'$.

7. Prove that an isometry is a collineation.

8. Prove that an isometry preserves betweenness.

9. Prove that an isometry preserves angle measure up to sign, i.e., an angle and its isometric image are congruent.

10. Prove that an isometry sends circles to circles.

11. Let $r > 0$. Prove that the transformation $\beta\left(\begin{bmatrix} x \\ y \end{bmatrix}\right) = \begin{bmatrix} rx \\ ry \end{bmatrix}$ is bijective.

12. Let $r > 0$. Prove that the transformation $\beta\left(\begin{bmatrix} x \\ y \end{bmatrix}\right) = \begin{bmatrix} rx \\ ry \end{bmatrix}$ is a similarity of ratio r.

13. If α is a similarity of ratio r and β is a similarity of ratio s, prove that $\alpha \circ \beta$ is a similarity of ratio rs.

14. Prove that a similarity of ratio r is a bijective transformation.

15. If α is a similarity of ratio r, prove that α^{-1} is a similarity of ratio $\frac{1}{r}$.

16. Let B, C, and D be distinct points on a circle centered at A. If $\square ABDC$ is a crossed quadrilateral, prove that $m\angle BAC \equiv 2m\angle BDC$.

17. Prove that the angle bisectors of a triangle are concurrent at a point equidistant from the three sides. Thus every triangle has an inscribed circle, called the *incircle;* its center point P is called the *incenter* of the triangle.

3.3   Appendix: Proof of Surjectivity

In this appendix we prove the fact that isometries are surjective.

Theorem 96 *Isometries are surjective.*

Proof. Given an isometry α and an arbitrary point A, let $A' = \alpha(A)$. If $A' = A$, we're done, so assume $A' \neq A$ and consider an equilateral triangle $\triangle ABC$ with sides of length AA'. Let $B' = \alpha(B)$ and $C' = \alpha(C)$. Again, if $B' = A$ or $C' = A$, we're done, so assume $B' \neq A$ and $C' \neq A$. Then $A'B' = AB$, $B'C' = BC$, and $C'A' = CA$ (α is an isometry), and $\triangle A'B'C' \cong \triangle ABC$ by SSS. Consider $\triangle AB'C'$, which is non-degenerate since A, B', and C' are distinct points on circle A'_A, and construct the point D on A_B such that $\angle BCD \cong \angle B'C'A$. Now $m\angle BAC = \pm 60$ and Proposition 86 implies $2m\angle BDC \equiv m\angle BAC = \pm m\angle B'A'C' \equiv \pm 2m\angle B'AC'$ so that $m\angle BDC = \pm 30 = \pm m\angle B'AC'$. Thus $\angle BDC \cong \angle B'AC'$ and $\triangle BCD \cong \triangle B'C'A$ by AAS (see Figure 3.3).

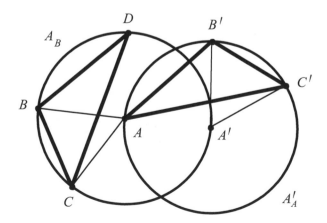

Figure 3.3. $\triangle BCD \cong \triangle B'C'A$.

Now $D' = \alpha(D)$ is on A'_A since $A'A = AD = A'D'$; D' is on B'_A since $B'A = BD = B'D'$ (CPCTC); and D' is on C'_A since $C'A = CD = C'D'$ (CPCTC). Thus A and D' are points of concurrency for circles A'_A, B'_A, and C'_A with non-collinear centers. Therefore $D' = A$ by uniqueness in Proposition 88 (see Figure 3.4). ∎

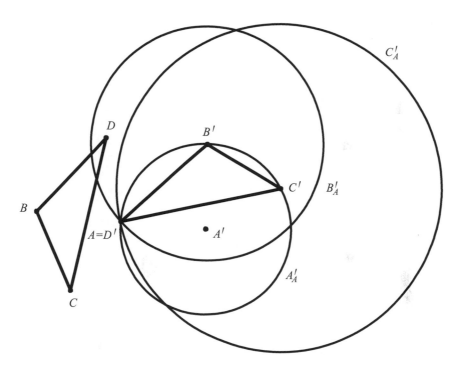

Figure 3.4. Circles A'_A, B'_A, and C'_A are concurrent at A and D'.

Chapter 4

Translations, Rotations, and Reflections

In grades 9–12 all students should apply transformations and use symmetry to...understand and represent translations, reflections, rotations, and dilations of objects in the plane by using sketches, coordinates, vectors, function notation, and matrices.

Principles and Standards for School Mathematics
National Council of Teachers of Mathematics (2000)

Studying geometric transformations provides opportunities for learners to describe patterns, discover basic features of isometries, make generalizations, and develop spatial competencies.

H. Bahadir Yanick, Math. Educator
Anadolu University

In this chapter we consider three important families of isometries and investigate some of their properties. We motivate each section with an exploratory activity. The instructions for these activities are generic and can be performed using any software package that supports geometric constructions. The Geometer's Sketchpad commands required by these activities appear in the appendix at the end of the chapter.

4.1 Translations

A translation of the plane is a transformation that slides the plane a finite distance in some direction. Exploratory activity 1, which follows below, uses the vector notation in the following definition:

Definition 97 *A **vector*** $\mathbf{v} = \begin{bmatrix} a \\ b \end{bmatrix}$ *is a quantity with norm (or magnitude)* $\|\mathbf{v}\| := \sqrt{a^2 + b^2}$ *and direction* Θ *defined by the equations* $\|\mathbf{v}\| \cos \Theta = a$ *and* $\|\mathbf{v}\| \sin \Theta = b$. *The values* a *and* b *are called the* x-***component*** *and* y-***component*** *of* $\mathbf{v}$, *respectively. The vector* $\mathbf{0} = \begin{bmatrix} 0 \\ 0 \end{bmatrix}$, *called the **zero vector**, has magnitude* 0 *and arbitrary direction. Given vectors* $\mathbf{v} = \begin{bmatrix} a \\ b \end{bmatrix}$ *and* $\mathbf{w} = \begin{bmatrix} c \\ d \end{bmatrix}$, *the* ***dot product*** *is defined to be* $\mathbf{v} \cdot \mathbf{w} := ac + bd$. *Thus* $\mathbf{v} \cdot \mathbf{v} = a^2 + b^2 = \|\mathbf{v}\|^2$. *If* $P = \begin{bmatrix} x \\ y \end{bmatrix}$ *and* $Q = \begin{bmatrix} x' \\ y' \end{bmatrix}$ *are points, the vector* $\mathbf{PQ} = \begin{bmatrix} x'-x \\ y'-y \end{bmatrix}$ *is called the **position vector** from* P *to* Q, *and* P *and* Q *are called the **initial** and **terminal** points of* $\mathbf{PQ}$. *A vector whose initial point is the origin* O *is in **standard position**.*

Graphically, we represent a position vector $\mathbf{PQ}$ as an arrow in the plane. Since a vector $\mathbf{v} = \begin{bmatrix} a \\ b \end{bmatrix}$ is completely determined by its components, we are free to position $\mathbf{v}$ anywhere in the plane we wish. However, once the initial point $P = \begin{bmatrix} x \\ y \end{bmatrix}$ has been established, the terminal point is uniquely determined: $Q = \begin{bmatrix} x+a \\ y+b \end{bmatrix}$.

Exploratory Activity 1: The Action of a Translation.

1. Construct two distinct points and label them P and Q.

2. Select P then Q, in that order.

3. Mark vector $\mathbf{PQ}$.

4. Construct line $\overleftrightarrow{PQ}$ and a point A on $\overleftrightarrow{PQ}$ such that $P - A - Q$.

5. Select point A, translate A by $\mathbf{PQ}$, and label the image A'.

6. Construct a point B off line $\overleftrightarrow{PQ}$.

7. Select point B, translate B by $\mathbf{PQ}$, and label the image B'.

8. Construct $\square PQB'B$. Measure $\overline{PQ}$, $\overline{BB'}$, $\overline{PB}$, $\overline{QB'}$, $\angle APB$, $\angle A'QB'$, and $\angle QB'B$. What special kind of quadrilateral is this? Why?

9. Construct $\square AA'B'B$. Measure $\overline{AA'}$, $\overline{BB'}$, $\overline{A'B'}$, $\overline{AB}$, $\angle ABB'$, $\angle BAP$, and $\angle B'A'Q$. What special kind of quadrilateral is this? Why?

10. Prove that $AB = A'B'$.

These observations indicate that *if* A' *is the image of a point* A *under the translation by vector* $\mathbf{PQ}$, *then* $\mathbf{AA'} = \mathbf{PQ}$. This motivates the definition of a translation.

Definition 98 *Let* P *and* Q *be points. The **translation from** P **to** Q *is the transformation* $\tau_{P,Q} : \mathbb{R}^2 \to \mathbb{R}^2$ *with the following properties:*

  1. $Q = \tau_{P,Q}(P)$.

2. If $P = Q$, then $\tau_{P,Q} = \iota$.

3. If $P \neq Q$, let A be any point on $\overleftrightarrow{PQ}$ and let B be any point off $\overleftrightarrow{PQ}$; let $A' = \tau_{P,Q}(A)$ and let $B' = \tau_{P,Q}(B)$. Then quadrilaterals $\square PQB'B$ and $\square AA'B'B$ are parallelograms (see Figure 4.1).

Let $\mathbf{v}$ be a vector, and let P and Q be points such that $\mathbf{PQ} = \mathbf{v}$. The *translation by vector* $\mathbf{v}$ is the transformation $\tau_{\mathbf{v}} := \tau_{P,Q}$. The *translation vector* of $\tau_{\mathbf{v}}$ is the vector $\mathbf{v}$ and the *length of* $\tau_{\mathbf{v}}$, denoted by $\|\tau_{\mathbf{v}}\|$, is the norm $\|\mathbf{v}\|$.

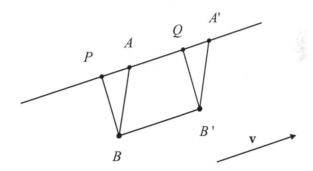

Figure 4.1. $\square PQB'B$ and $\square AA'B'B$ are parallelograms.

A translation can be expressed in terms of a "vector sum" (see Proposition 102 below).

Definition 99 *Let $a \in \mathbb{R}$ and let $\mathbf{u} = \begin{bmatrix} u_1 \\ u_2 \end{bmatrix}$ and $\mathbf{v} = \begin{bmatrix} v_1 \\ v_2 \end{bmatrix}$ be vectors. Define the scalar product*

$$a\mathbf{u} := \begin{bmatrix} au_1 \\ au_2 \end{bmatrix},$$

the vector sum

$$\mathbf{u} + \mathbf{v} := \begin{bmatrix} u_1 + v_1 \\ u_2 + v_2 \end{bmatrix},$$

and the vector difference

$$\mathbf{u} - \mathbf{v} := \mathbf{u} + (-1)\mathbf{v} = \begin{bmatrix} u_1 - v_1 \\ u_2 - v_2 \end{bmatrix}.$$

To picture a vector sum, position $\mathbf{PQ}$ with its terminal point at the initial point of $\mathbf{RS}$. Then $\mathbf{PQ}$ and $\mathbf{RS}$ determine a parallelogram whose diagonal terminating at S represents $\mathbf{PQ} + \mathbf{RS}$ (see Figure 4.2). In particular, if O is the origin, then $\mathbf{OP} + \mathbf{PQ} = \mathbf{OQ}$ and equivalently, $\mathbf{PQ} = \mathbf{OQ} - \mathbf{OP}$. Note that by positioning $\mathbf{RS}$ with its terminal point at the initial point of $\mathbf{PQ}$, we obtain a parallelogram whose diagonal terminating at Q represents $\mathbf{RS} + \mathbf{PQ}$. Since the diagonals $\mathbf{PQ} + \mathbf{RS}$ and $\mathbf{RS} + \mathbf{PQ}$ represent the same vector, $\mathbf{PQ} + \mathbf{RS} = \mathbf{RS} + \mathbf{PQ}$ and vector addition is commutative.

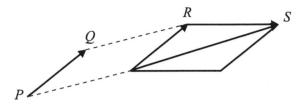

Figure 4.2. Vector sum of **PQ** and **RS**.

Now identify points in $\mathbb{R}^2$ with terminal points of vectors in standard position. Then a transformation ϕ sends a vector **OP** to the vector **O′P′**, where $O' = \phi(O)$ and $P' = \phi(P)$.

Definition 100 *A transformation* $\phi : \mathbb{R}^2 \to \mathbb{R}^2$ *is **linear** if for all vectors* $\begin{bmatrix} s \\ t \end{bmatrix}$ *and* $\begin{bmatrix} u \\ v \end{bmatrix}$, $a, b \in \mathbb{R}$,

1. $\phi\left(a\begin{bmatrix} s \\ t \end{bmatrix}\right) = a\phi\left(\begin{bmatrix} s \\ t \end{bmatrix}\right)$ *(*ϕ *is **homogeneous**), and*

2. $\phi\left(\begin{bmatrix} s \\ t \end{bmatrix} + \begin{bmatrix} u \\ v \end{bmatrix}\right) = \phi\left(\begin{bmatrix} s \\ t \end{bmatrix}\right) + \phi\left(\begin{bmatrix} u \\ v \end{bmatrix}\right)$ *(*ϕ *is **additive**).*

Together, properties (1) and (2) are equivalent to the single statement

$$\phi\left(a\begin{bmatrix} s \\ t \end{bmatrix} + b\begin{bmatrix} u \\ v \end{bmatrix}\right) = a\phi\left(\begin{bmatrix} s \\ t \end{bmatrix}\right) + b\phi\left(\begin{bmatrix} u \\ v \end{bmatrix}\right).$$

Proposition 101 *An isometry is linear if and only if it fixes the origin.*

Proof. The proof is left as an exercise to the reader. ∎

Proposition 102 *Let* **v** *be a vector, let* O *be the origin, and let* P *be a point. Then* $Q = \tau_{\mathbf{v}}(P)$ *if and only if* **OQ** = **OP** + **v**.

Proof. By definition of translation by vector **v**, $Q = \tau_{\mathbf{v}}(P)$ if and only if **v** = **PQ** = **OQ** − **OP** if and only if **OQ** = **OP** + **v**. ∎

A translation $\tau_{\mathbf{v}}$ acts on a vector **w** as the identity, i.e., $\tau_{\mathbf{v}}(\mathbf{w}) = \mathbf{w}$, which seems intuitively obvious since we use the same vector **v** to translate the initial and terminal points of a position vector. But to be more precise, let R and S be points, let $R' = \tau_{\mathbf{v}}(R)$, and let $S' = \tau_{\mathbf{v}}(S)$. Then by Proposition 102 we have

$$\mathbf{R'S'} = \mathbf{OS'} - \mathbf{OR'} = (\mathbf{OS} + \mathbf{v}) - (\mathbf{OR} + \mathbf{v}) = \mathbf{OS} - \mathbf{OR} = \mathbf{RS}.$$

Therefore $\tau_{\mathbf{v}}(\mathbf{RS}) = \mathbf{R'S'} = \mathbf{RS}$.

Theorem 103 *Translations are isometries.*

Proof. Let $\mathbf{v}$ be a vector. Given points P and Q, let $P' = \tau_{\mathbf{v}}(P)$ and $Q' = \tau_{\mathbf{v}}(Q)$. Then by the remark above, $\tau_{\mathbf{v}}(\mathbf{PQ}) = \mathbf{P'Q'} = \mathbf{PQ}$ so that $PQ = \|\mathbf{PQ}\| = \|\mathbf{P'Q'}\| = P'Q'$. Therefore $\tau_{\mathbf{v}}$ is an isometry. $\blacksquare$

By definition, a translation is uniquely determined by a point and its image. Consequently, we shall often refer to a general translation τ without specific reference to a point P and its image Q or to the vector $\mathbf{v} = \mathbf{PQ}$. When we need the vector $\mathbf{v}$, for example, we simply evaluate τ at any point $\left[\begin{smallmatrix} x \\ y \end{smallmatrix}\right]$, obtain the image point $\left[\begin{smallmatrix} x' \\ y' \end{smallmatrix}\right]$, then compute the components $a = x' - x$ and $b = y' - y$ of $\mathbf{v}$.

The *equations* of a transformation $\alpha : \mathbb{R}^2 \to \mathbb{R}^2$ are the *coordinate functions* $x' = \alpha_1(x, y)$ and $y' = \alpha_2(x, y)$ such that

$$\begin{bmatrix} x' \\ y' \end{bmatrix} = \alpha\left(\begin{bmatrix} x \\ y \end{bmatrix} \right).$$

We evaluate α at the point $X = \left[\begin{smallmatrix} x \\ y \end{smallmatrix}\right]$ as we did in Chapter 3, by evaluating the coordinate functions x' and y' at (x, y). Thus if $\mathbf{v} = \left[\begin{smallmatrix} a \\ b \end{smallmatrix}\right]$ is a vector, and X and $X' = \tau_{\mathbf{v}}(X)$ are identified with the position vectors $\mathbf{OX}$ and $\mathbf{OX'}$, Proposition 102 gives

$$\begin{bmatrix} x' \\ y' \end{bmatrix} = \mathbf{OX'} = \mathbf{OX} + \mathbf{v} = \begin{bmatrix} x \\ y \end{bmatrix} + \begin{bmatrix} a \\ b \end{bmatrix} = \begin{bmatrix} x + a \\ y + b \end{bmatrix},$$

and by equating components we obtain:

Proposition 104 *Let* $\mathbf{v} = \left[\begin{smallmatrix} a \\ b \end{smallmatrix}\right]$ *be a vector. The equations of* $\tau_{\mathbf{v}}$ *are*

$$\begin{aligned} x' &= x + a \\ y' &= y + b. \end{aligned}$$

Example 105 Let $P = \left[\begin{smallmatrix} 4 \\ 5 \end{smallmatrix}\right]$ and $Q = \left[\begin{smallmatrix} -1 \\ 3 \end{smallmatrix}\right]$. Then $\mathbf{PQ} = \left[\begin{smallmatrix} -5 \\ -2 \end{smallmatrix}\right]$ and the equations for $\tau_{\mathbf{PQ}}$ are

$$\begin{aligned} x' &= x - 5 \\ y' &= y - 2. \end{aligned}$$

In particular, $\tau_{\mathbf{PQ}}\left(\left[\begin{smallmatrix} 7 \\ -5 \end{smallmatrix}\right] \right) = \left[\begin{smallmatrix} 2 \\ -7 \end{smallmatrix}\right]$.

Although function composition is not commutative in general, the composition of translations is commutative. Intuitively, commutativity tells us that either of the two routes along the edges of a parallelogram from one vertex to its diagonal opposite leads to the same destination. This fact is part 2 of the next proposition.

Proposition 106 *Let* **v** *and* **w** *be vectors.*

1. *The composition of translations is a translation. In fact,*

$$\tau_{\mathbf{w}} \circ \tau_{\mathbf{v}} = \tau_{\mathbf{v}+\mathbf{w}}.$$

2. *The composition of translations commutes, i.e.,*

$$\tau_{\mathbf{v}} \circ \tau_{\mathbf{w}} = \tau_{\mathbf{w}} \circ \tau_{\mathbf{v}}.$$

3. *The inverse of a translation is a translation. In fact,*

$$\tau_{\mathbf{v}}^{-1} = \tau_{-\mathbf{v}}.$$

Proof. (1) Let A be a point, let $A' = \tau_{\mathbf{v}}(A)$, and let $A'' = \tau_{\mathbf{w}}(A')$. Then $A'' = \tau_{\mathbf{w}}(\tau_{\mathbf{v}}(A)) = (\tau_{\mathbf{w}} \circ \tau_{\mathbf{v}})(A)$. On the other hand, by Proposition 102 we have

$$\mathbf{OA''} = \mathbf{OA'} + \mathbf{w} = (\mathbf{OA} + \mathbf{v}) + \mathbf{w} = \mathbf{OA} + (\mathbf{v} + \mathbf{w})$$

so that $A'' = \tau_{\mathbf{v}+\mathbf{w}}(A)$. Therefore $\tau_{\mathbf{w}} \circ \tau_{\mathbf{v}} = \tau_{\mathbf{v}+\mathbf{w}}$.

(2) By part 1 and the fact that vector addition is commutative, we have

$$\tau_{\mathbf{v}} \circ \tau_{\mathbf{w}} = \tau_{\mathbf{w}+\mathbf{v}} = \tau_{\mathbf{v}+\mathbf{w}} = \tau_{\mathbf{w}} \circ \tau_{\mathbf{v}}.$$

(3) The proof is left as an exercise for the reader. ∎

Definition 107 *A collineation α is a **dilatation** if $\alpha(l) \parallel l$ for every line l.*

Theorem 108 *Translations are dilatations.*

Proof. Let τ be a translation. Then τ is an isometry by Theorem 103. Let l be a line, let A and B be distinct points on line l, and let $A' = \tau(A)$ and $B' = \tau(B)$ (see Figure 4.3).

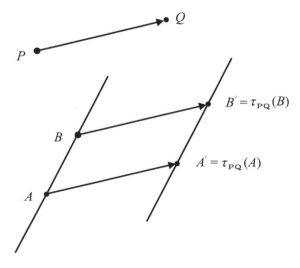

Figure 4.3. Proof of Theorem 108.

Then $\tau(l) = \overleftrightarrow{A'B'}$ since τ is a collineation by Proposition 92, part 1, and $\tau = \tau_{\mathbf{AA'}} = \tau_{\mathbf{BB'}}$ by definition of translation. If A, B, A', and B' are collinear, then $\overleftrightarrow{AB} = \overleftrightarrow{A'B'}$ so that $\overleftrightarrow{AB} \parallel \overleftrightarrow{A'B'}$. If A, B, A', and B' are non-collinear, then B lies off $\overleftrightarrow{AA'}$ and $\square AA'B'B$ is a parallelogram since $\mathbf{AA'} = \mathbf{BB'}$, and it follows that $\overleftrightarrow{AB} \parallel \overleftrightarrow{A'B'}$. ∎

We conclude this section with a key definition that will allow us to distinguish between the various families of isometries.

Definition 109 *A transformation α **fixes** a set s if $\alpha(s) = s$. A transformation α fixes a set s **pointwise** if $\alpha(P) = P$ for each point $P \in s$.*

Theorem 110 *Let P and Q be distinct points. The translation $\tau_{\mathbf{PQ}}$ is fixed point free, but fixes every line parallel to $\overleftrightarrow{PQ}$.*

Proof. Let A be a point and let $A' = \tau_{\mathbf{PQ}}(A)$. Then $A \neq A'$ since $\mathbf{AA'} = \mathbf{PQ} \neq \mathbf{0}$, and $\tau_{\mathbf{PQ}}$ is fixed point free. Furthermore, let l be a line parallel to $\overleftrightarrow{PQ}$ and let B be a point on l. Then $\overleftrightarrow{BB'} \parallel \overleftrightarrow{PQ} \parallel l$ and $B \in \overleftrightarrow{BB'} \cap l$ implies $\overleftrightarrow{BB'} = l$. Therefore B' is on l and $\tau_{\mathbf{PQ}}(l) = l$ since translations are collineations. ∎

Exercises

1. Let τ be the translation such that $\tau\left(\begin{bmatrix} -1 \\ 3 \end{bmatrix}\right) = \begin{bmatrix} 5 \\ 2 \end{bmatrix}$.

   (a) Find the vector of τ.

   (b) Find the equations of τ.

   (c) Find $\tau\left(\begin{bmatrix} 0 \\ 0 \end{bmatrix}\right)$, $\tau\left(\begin{bmatrix} 3 \\ -7 \end{bmatrix}\right)$, and $\tau\left(\begin{bmatrix} -5 \\ -2 \end{bmatrix}\right)$.

   (d) Find x and y such that $\tau\left(\begin{bmatrix} x \\ y \end{bmatrix}\right) = \begin{bmatrix} 0 \\ 0 \end{bmatrix}$.

2. Let τ be the translation such that $\tau\left(\begin{bmatrix} 4 \\ 6 \end{bmatrix}\right) = \begin{bmatrix} 7 \\ 10 \end{bmatrix}$.

   (a) Find the vector of τ.

   (b) Find the equations of τ.

   (c) Find $\tau\left(\begin{bmatrix} 0 \\ 0 \end{bmatrix}\right)$, $\tau\left(\begin{bmatrix} 1 \\ 2 \end{bmatrix}\right)$, and $\tau\left(\begin{bmatrix} -3 \\ -4 \end{bmatrix}\right)$.

   (d) Find x and y such that $\tau\left(\begin{bmatrix} x \\ y \end{bmatrix}\right) = \begin{bmatrix} 0 \\ 0 \end{bmatrix}$.

3. Let $P = \begin{bmatrix} 4 \\ -1 \end{bmatrix}$ and $Q = \begin{bmatrix} -3 \\ 5 \end{bmatrix}$.

   (a) Find the vector of $\tau_{P,Q}$.

(b) Find the equations of $\tau_{P,Q}$.

(c) Find $\tau_{P,Q}\left(\begin{bmatrix} 3 \\ 6 \end{bmatrix}\right)$, $\tau_{P,Q}\left(\begin{bmatrix} 1 \\ 2 \end{bmatrix}\right)$, and $\tau_{P,Q}\left(\begin{bmatrix} -3 \\ -4 \end{bmatrix}\right)$.

(d) Let l be the line with equation $2X + 3Y + 4 = 0$. Find the equation of the line $l' = \tau_{P,Q}(l)$.

4. Prove that $\tau_{\mathbf{v}}^{-1} = \tau_{-\mathbf{v}}$.

5. Prove that an isometry is linear if and only if it fixes the origin.

6. Is a non-trivial translation linear? Explain.

4.2   Rotations

In this section we define rotations, derive their equations, and investigate some of their fundamental properties.

Exploratory Activity 2: The Action of a Rotation.

1. Construct two distinct points and label them C and P.

2. Construct circle C_P.

3. Construct ray $\overrightarrow{CP}$.

4. Rotate ray $\overrightarrow{CP}$ about C through an angle of $72°$.

5. Construct the intersection point of C_P and the image of $\overrightarrow{CP}$ in Step 4, and label it P' (this is the image of P under a $72°$ rotation about C).

6. Measure angle $\angle PCP'$.

7. Construct a point outside circle C_P and label it Q.

8. Following steps 2–5 above, construct the image of Q under a $72°$ rotation about C and label it Q'.

9. Measure angle $\angle QCQ'$.

10. Construct segments $\overline{PQ}$ and $\overline{P'Q'}$, and measure their lengths. What do you observe?

11. Prove that $PQ = P'Q'$.

This activity demonstrates that the image of a point X under a 72° rotation about C is the point X' such that $CX = CX'$ and $m\angle XCX' = 72°$. This motivates the definition of a rotation (Definition 111). Furthermore, your proof in Step 11 essentially proves that rotations are isometries.

Definition 111 *Let C and P be points, and let $\Theta \in \mathbb{R}$. The **rotation about** C **of** $\Theta°$ is the transformation $\rho_{C,\Theta} : \mathbb{R}^2 \to \mathbb{R}^2$ with the following properties:*

1. $\rho_{C,\Theta}(C) = C$.

2. *If $P \neq C$ and $P' = \rho_{C,\Theta}(P)$, then $CP' = CP$ and $m\angle PCP' \equiv \Theta$.*

*The point C is the **center** of rotation $\rho_{C,\Theta}$, and Θ is the **angle** of rotation (see Figure 4.4).*

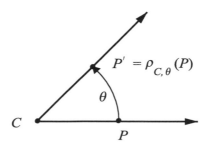

Figure 4.4. Definition of rotation $\rho_{C,\Theta}$.

Of course, $\rho_{C,\Theta_1} = \rho_{C,\Theta_2}$ if and only if $\Theta_1 \equiv \Theta_2$.

Theorem 112 *Rotations are isometries.*

Proof. Let $\rho_{C,\Theta}$ be a rotation. Let C, P, and Q be points with P and Q distinct; let $P' = \rho_{C,\Theta}(P)$ and $Q' = \rho_{C,\Theta}(Q)$. If $P = C$, then by definition, $PQ = CQ = CQ' = P'Q'$, and similarly for $Q = C$. So assume that C, P, and Q are distinct. If C, P, and Q are non-collinear, then $\triangle PCQ \cong \triangle P'CQ'$ by SAS, and $PQ = P'Q'$ by CPCTC. If C, P, and Q are collinear with $C-P-Q$, then $PQ = CQ - CP = CQ' - CP' = P'Q'$, since $CP = CP'$ and $CQ = CQ'$ by definition, and similarly for $C-Q-P$. But if $P-C-Q$, then $P'-C-Q'$ since $m\angle PCP' = m\angle QCQ' \equiv \Theta$. Therefore $PQ = CP + CQ = CP' + CQ' = P'Q'$. ∎

Before we derive the equations of a general rotation, we consider the special case of rotations $\rho_{O,\Theta}$ about the origin. Since $\rho_{O,\Theta}$ is an isometry that fixes O, it is linear by Proposition 101. Let $\mathbf{e}_1 = \begin{bmatrix} 1 \\ 0 \end{bmatrix}$ and $\mathbf{e}_2 = \begin{bmatrix} 0 \\ 1 \end{bmatrix}$ be the standard unit coordinate vectors; let $\mathbf{e}_1' = \rho_{O,\Theta}(\mathbf{e}_1)$ and $\mathbf{e}_2' = \rho_{O,\Theta}(\mathbf{e}_2)$. Then

$$\rho_{O,\Theta}\left(\begin{bmatrix} x \\ y \end{bmatrix}\right) = \rho_{O,\Theta}(x\mathbf{e}_1 + y\mathbf{e}_2) = x\rho_{O,\Theta}(\mathbf{e}_1) + y\rho_{O,\Theta}(\mathbf{e}_2) = x\mathbf{e}_1' + y\mathbf{e}_2'$$

and $\rho_{O,\Theta}$ is completely determined by its action on $\mathbf{e}_1$ and $\mathbf{e}_2$. Let E_1 and E_2 be the terminal points of $\mathbf{e}_1$ and $\mathbf{e}_2$ in standard position. Since $m\angle E_1OE_1' = \Theta$ we have

$$\mathbf{e}_1' = \begin{bmatrix} \cos\Theta \\ \sin\Theta \end{bmatrix}.$$

Furthermore, since $m\angle E_1OE_2 = 90$ and $m\angle E_2OE_2' = \Theta$, we have $m\angle E_1OE_2' = \Theta + 90$ and

$$\mathbf{e}_2' = \rho_{O,\Theta+90}\left(\mathbf{e}_1\right) = \begin{bmatrix} \cos\left(\Theta + 90\right) \\ \sin\left(\Theta + 90\right) \end{bmatrix} = \begin{bmatrix} -\sin\Theta \\ \cos\Theta \end{bmatrix}.$$

Consequently,

$$\begin{bmatrix} x' \\ y' \end{bmatrix} = x\mathbf{e}_1' + y\mathbf{e}_2' = \begin{bmatrix} x\cos\Theta \\ x\sin\Theta \end{bmatrix} + \begin{bmatrix} -y\sin\Theta \\ y\cos\Theta \end{bmatrix} = \begin{bmatrix} x\cos\Theta - y\sin\Theta \\ x\sin\Theta + y\cos\Theta \end{bmatrix},$$

and we have proved:

Theorem 113 *Let $\Theta \in \mathbb{R}$. The equations for $\rho_{O,\Theta}$ are*

$$\begin{aligned} x' &= x\cos\Theta - y\sin\Theta \\ y' &= x\sin\Theta + y\cos\Theta. \end{aligned} \tag{4.1}$$

Note that Equation (4.1) can be expressed in terms of matrix multiplication:

$$\begin{bmatrix} x' \\ y' \end{bmatrix} = \begin{bmatrix} \cos\Theta & -\sin\Theta \\ \sin\Theta & \cos\Theta \end{bmatrix} \begin{bmatrix} x \\ y \end{bmatrix}.$$

Thus the action of $\rho_{O,\Theta}$ on a vector $\mathbf{v} = \begin{bmatrix} x \\ y \end{bmatrix}$ is given by

$$\rho_{O,\Theta}\left(\mathbf{v}\right) = \begin{bmatrix} \cos\Theta & -\sin\Theta \\ \sin\Theta & \cos\Theta \end{bmatrix} \mathbf{v}.$$

Now think of a general rotation $\rho_{C,\Theta}$ about the point $C = \begin{bmatrix} a \\ b \end{bmatrix}$ as the composition of the following three isometries (see Figure 4.5):

1. Translate by vector $\mathbf{CO}$.

2. Rotate about O through an angle of $\Theta°$.

3. Translate by vector $\mathbf{OC}$.

Then

$$\rho_{C,\Theta} = \tau_{\mathbf{CO}} \circ \rho_{O,\Theta} \circ \tau_{\mathbf{CO}}^{-1}.$$

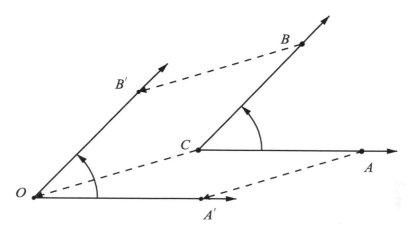

Figure 4.5. $\rho_{C,\Theta} = \tau_{\mathbf{CO}} \circ \rho_{O,\Theta} \circ \tau_{\mathbf{CO}}^{-1}$.

This particular composition, called the *conjugation of $\rho_{O,\Theta}$ by $\tau_{\mathbf{OC}}$*, is fundamentally important and will appear again in Chapter 6. Thus the equations of $\rho_{C,\Theta}$ are given by composing the equations of isometries (1)–(3) above:

$$\begin{bmatrix} x' \\ y' \end{bmatrix} = (\tau_{\mathbf{OC}} \circ \rho_{O,\Theta} \circ \tau_{\mathbf{CO}}) \left(\begin{bmatrix} x \\ y \end{bmatrix} \right) = \tau_{\mathbf{OC}} \left(\rho_{O,\Theta} \left(\begin{bmatrix} x - a \\ y - b \end{bmatrix} \right) \right)$$

$$= \tau_{\mathbf{OC}} \left(\begin{bmatrix} (x - a)\cos\Theta - (y - b)\sin\Theta \\ (x - a)\sin\Theta + (y - b)\cos\Theta \end{bmatrix} \right)$$

$$= \begin{bmatrix} (x - a)\cos\Theta - (y - b)\sin\Theta + a \\ (x - a)\sin\Theta + (y - b)\cos\Theta + b \end{bmatrix}.$$

Since translations act on vectors as the identity, a general rotation acts on a vector **v** by rotating **v** about the origin:

$$\rho_{C,\Theta}(\mathbf{v}) = \left(\tau_{\mathbf{CO}} \circ \rho_{O,\Theta} \circ \tau_{\mathbf{CO}}^{-1} \right)(\mathbf{v}) = \rho_{O,\Theta}(\mathbf{v}).$$

Thus the action of a rotation on a vector only depends on the rotation angle and *not* on the center of rotation. But for points we have:

Theorem 114 *Let $C = \begin{bmatrix} a \\ b \end{bmatrix}$ and let $\Theta \in \mathbb{R}$. The equations of $\rho_{C,\Theta}$ are*

$$x' = (x - a)\cos\Theta - (y - b)\sin\Theta + a$$
$$y' = (x - a)\sin\Theta + (y - b)\cos\Theta + b.$$

Example 115 *Let $C = \begin{bmatrix} 2 \\ -3 \end{bmatrix}$. Then the equations for $\rho_{C,90}$ are*

$$x' = (x - 2)\cos 90 - (y + 3)\sin 90 + 2 = -y - 1$$
$$y' = (x - 2)\sin 90 + (y + 3)\cos 90 - 3 = x - 5.$$

In particular, $\rho_{C,90}\left(\begin{bmatrix} 5 \\ -1 \end{bmatrix}\right) = \begin{bmatrix} 0 \\ 0 \end{bmatrix}$.

The proof of our next proposition is similar to the proof of Proposition 106 and is left as an exercise for the reader:

Proposition 116 *Let C be a point and let $\Theta, \Phi \in \mathbb{R}$.*

1. *A composition of rotations about C is a rotation. In fact,*

$$\rho_{C,\Theta} \circ \rho_{C,\Phi} = \rho_{C,\Theta+\Phi}.$$

2. *A composition of rotations about C commutes, i.e.,*

$$\rho_{C,\Theta} \circ \rho_{C,\Phi} = \rho_{C,\Phi} \circ \rho_{C,\Theta}.$$

3. *The inverse of a rotation about C is a rotation about C. In fact,*

$$\rho_{C,\Theta}^{-1} = \rho_{C,-\Theta}.$$

Proposition 117 *A non-trivial rotation $\rho_{C,\Theta}$ fixes every circle with center C and has exactly one fixed point, namely C.*

Proof. That $\rho_{C,\Theta}(C) = C$ follows by definition. From the equations of a rotation it is evident that $\rho_{C,\Theta} = \iota$ (the identity) if and only if $\Theta \equiv 0$, so $\Theta° \neq 0°$ by assumption. Let P be any point distinct from C, and let and $P' = \rho_{C,\Theta}(P)$; then $P \neq P'$ since $m\angle PCP' \notin 0°$. Therefore C is the unique point fixed by $\rho_{C,\Theta}$. Furthermore, let Q be any point on C_P and let $Q' = \rho_{C,\Theta}(Q)$. Then by definition $CP = CQ = CQ'$ and Q' is on C_P. Hence $\rho_{C,\Theta}(C_P) \subseteq C_P$. Since $\rho_{C,\Theta}$ is an isometry, $\rho_{C,\Theta}(C_P) = C_P$ by Proposition 92, part 4. ∎

Since a non-trivial rotation has exactly one fixed point and a non-trivial translation has none, we have:

Corollary 118 *A non-trivial rotation is not a translation.*

Rotations of 180°, called *halfturns*, play an important theoretical role in this exposition. We conclude this section with a brief look at some of their special properties.

Definition 119 *Let C be a point. The **halfturn about** C is the transformation $\varphi_C : \mathbb{R}^2 \to \mathbb{R}^2$ defined by $\varphi_C := \rho_{C,180}$. The point C is the **center** of halfturn φ_C.*

Halfturns are characterized in the following way:

Proposition 120 *Let C be a point and let $\alpha : \mathbb{R}^2 \to \mathbb{R}^2$ be a transformation. Then $\alpha = \varphi_C$ if and only if*

1. $\alpha(C) = C$.

2. If P is a point distinct from C and $P' = \alpha(P)$, then C is the midpoint of $\overline{PP'}$.

Proof. ($\Rightarrow$) Assume $\alpha = \varphi_C$. Then $\alpha(C) = \varphi_C(C) = \rho_{C,180}(C) = C$ and Condition (1) holds. Furthermore, let P be a point distinct from C and let $P' = \alpha(P) = \varphi_C(P) = \rho_{C,180}(P)$. Then $PC = CP'$ and $m\angle PCP' = 180$ by Definition 111 so that C is the midpoint of $\overline{PP'}$, and Condition (2) holds.

($\Leftarrow$) Assume Conditions (1) and (2) hold. Then $\alpha(C) = C = \rho_{C,180}(C)$, and if P is a point distinct from C and $P' = \alpha(P)$, then $CP' = CP$ and $m\angle PCP' = 180$ so that $\alpha(P) = P' = \rho_{C,180}(P)$ by Definition 111. Therefore $\alpha = \rho_{C,180} = \varphi_C$. ∎

Some authors refer to the halfturn about C as *"the reflection in point C."* However, in this exposition we use the "halfturn" terminology exclusively. The equations of the halfturn about $C = \begin{bmatrix} a \\ b \end{bmatrix}$ follow immediately from Theorem 114:
$$x' = (x - a)\cos 180 - (y - b)\sin 180 + a = 2a - x$$
$$y' = (x - a)\sin 180 + (y - b)\cos 180 + b = 2b - y.$$

Corollary 121 *If $C = \begin{bmatrix} a \\ b \end{bmatrix}$, the equations of φ_C are*
$$x' = 2a - x$$
$$y' = 2b - y.$$

Example 122 Let O denote the origin; the equations for the halfturn about the origin φ_O are
$$x' = -x$$
$$y' = -y.$$

Recall that a function $f : \mathbb{R} \to \mathbb{R}$ is odd if $f(-x) = -f(x)$. An example of such a function is $f(x) = \sin(x)$. Let f be odd and consider a point $P = \begin{bmatrix} x \\ f(x) \end{bmatrix}$ on the graph of f. The image of P under the halfturn φ_O is
$$\varphi_O\left(\begin{bmatrix} x \\ f(x) \end{bmatrix}\right) = \begin{bmatrix} -x \\ -f(x) \end{bmatrix} = \begin{bmatrix} -x \\ f(-x) \end{bmatrix},$$
which is also a point on the graph of f. Thus φ_O fixes the graph of f.

Proposition 123 *A line l is fixed by φ_C if and only if C is on l.*

Proof. Let φ_C be a halfturn and let l be a line.

($\Leftarrow$) If C is on l, consider a point P on l distinct from C. Let $P' = \varphi_C(P)$; by definition, C is the midpoint of $\overline{PP'}$. Hence P' is on l and l is fixed by φ_C.

($\Rightarrow$) If C is off l, consider any point P on l. Then $P' = \varphi_C(P)$ is off l since otherwise the midpoint of $\overline{PP'}$, which is C, would be on l. Therefore l is not fixed by φ_C. ∎

Definition 124 *A non-trivial transformation α is an **involution** if $\alpha^2 = \iota$.*

Note that an involution α has the property that $\alpha^{-1} = \alpha$.

Proposition 125 *A halfturn is both an involution and a dilatation.*

Proof. The proof is left as an exercise for the reader. ■

Corollary 126 *Involutory rotations are halfturns.*

Proof. An involutory rotation $\rho_{C,\Theta}$ is a non-trivial rotation such that $\rho_{C,\Theta}^2 = \iota = \rho_{C,0}$. By Proposition 116, $\rho_{C,\Theta}^2 = \rho_{C,2\Theta}$ so that $\rho_{C,2\Theta} = \rho_{C,0}$ and $2\Theta \equiv 0$. Hence there exists some $k \in \mathbb{Z}$, such that $2\Theta = 360k$; consequently $\Theta = 180k$. Now k cannot be even since $\rho_{C,\Theta} \neq \iota$ implies that Θ is not a multiple of 360. Therefore $k = 2n + 1$ is odd and $\Theta = 180 + 360n$. It follows that $\Theta \equiv 180$ and $\rho_{C,\Theta} = \varphi_C$ as claimed. ■

Exercises

1. Find the coordinates of the point $\rho_{O,30}\left(\begin{bmatrix} 3 \\ 6 \end{bmatrix}\right)$.

2. Let $Q = \begin{bmatrix} -3 \\ 5 \end{bmatrix}$. Find the coordinates of the point $\rho_{Q,45}\left(\begin{bmatrix} 3 \\ 6 \end{bmatrix}\right)$.

3. Let l be the line with equation $2X + 3Y + 4 = 0$.

   (a) Find the equation of the line $\rho_{O,30}(l)$.
   (b) Let $Q = \begin{bmatrix} -3 \\ 5 \end{bmatrix}$. Find the equation of the line $\rho_{Q,45}(l)$.

4. Let C be a point and let $\Theta, \Phi \in \mathbb{R}$. Prove that $\rho_{C,\Theta} \circ \rho_{C,\Phi} = \rho_{C,\Theta+\Phi}$.

5. Let C be a point and let $\Theta, \Phi \in \mathbb{R}$. Prove that $\rho_{C,\Theta} \circ \rho_{C,\Phi} = \rho_{C,\Phi} \circ \rho_{C,\Theta}$.

6. Let C be a point and let $\Theta \in \mathbb{R}$. Prove that $\rho_{C,\Theta}^{-1} = \rho_{C,-\Theta}$.

7. People in distress on a deserted island sometimes write SOS in the sand.

   (a) Why is this signal particularly effective when viewed from searching aircraft?
   (b) The word SWIMS, like SOS, reads the same after performing a halfturn about its centroid. Find at least five other words that are preserved under a halfturn about their centroids.

8. Try it! Plot the graph of $y = \sin(x)$ on graph paper, pierce the graph paper at the origin with your compass point and push the compass point into your writing surface. This provides a point around which you can rotate your graph paper. Now physically rotate your graph paper $180°$ and observe that the graph of $y = \sin(x)$ is fixed by φ_O in the sense defined above.

9. Find the coordinates for the center of the halfturn whose equations are $x' = -x + 3$ and $y' = -y - 8$.

10. Let $P = \begin{bmatrix} 2 \\ 3 \end{bmatrix}$; let l be the line with equation $5X - Y + 7 = 0$.

    (a) Find the equations of φ_P.
    (b) Find the image of $\begin{bmatrix} 1 \\ 2 \end{bmatrix}$ and $\begin{bmatrix} -2 \\ 5 \end{bmatrix}$ under φ_P.
    (c) Find the equation of the line $l' = \varphi_P(l)$. On graph paper, plot point P and draw lines l and l'.

11. Repeat Exercise 10 with $P = \begin{bmatrix} -3 \\ 2 \end{bmatrix}$.

12. $P = \begin{bmatrix} 3 \\ -2 \end{bmatrix}$ and $Q = \begin{bmatrix} -5 \\ 7 \end{bmatrix}$. Find the equations of the composition $\varphi_Q \circ \varphi_P$ and inspect them carefully. These are the equations of an isometry we discussed earlier in the course. Can you identify which?

13. Let $P = \begin{bmatrix} a \\ b \end{bmatrix}$ and $Q = \begin{bmatrix} c \\ d \end{bmatrix}$ be distinct points. Find the equations of $\varphi_Q \circ \varphi_P$ and prove that the composition $\varphi_Q \circ \varphi_P$ is a translation τ. Find the vector of τ.

14. In the diagram below, circles A_P and B_P intersect at points P and Q. Use a MIRA to find a line through P distinct from $\overleftrightarrow{PQ}$ that intersects circles A_P and B_P in chords of equal length.

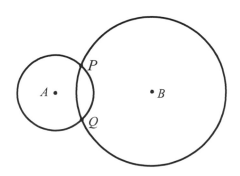

15. For any point P, prove that $\varphi_P^{-1} = \varphi_P$.

16. Prove that $\varphi_A \circ \varphi_B = \varphi_B \circ \varphi_A$ if and only if $A = B$.

17. Prove Proposition 125: A halfturn is both an involution and a dilatation.

18. Which rotations are linear? Explain.

4.3   Reflections

The reflection of the plane in a line m sends each point P to its mirror image P' with m acting as a mirror.

Exploratory Activity 3: The Action of a Reflection.

1. Open a new sketch, construct a line, and label it m.

2. Mark line m as a mirror.

3. Construct a point off line m and label it P.

4. Reflect point P in line m and label the image point P'.

5. Construct segment $\overline{PP'}$ and the point of intersection $m \cap \overline{PP'}$. Label the intersection point M.

6. Measure the distances MP and MP'. What do you observe?

7. Construct a point on line m distinct from M and label it A.

8. Measure $\angle AMP$. What special kind of angle is this?

9. Construct a point off line m distinct from P and on the same side of m as P, and label it Q.

10. Reflect point Q in line m and label the image point Q'.

11. Measure the distances PQ and $P'Q'$. What do you observe?

12. Prove that $PQ = P'Q'$.

These observations indicate that the mirror line m is the perpendicular bisector of the segment connecting the point P and its mirror image P'. This motivates the definition of a reflection (Definition 127). Furthermore, your proof in Step 12 essentially proves that reflections are isometries.

Definition 127 *Let m be a line. The **reflection in line** m is the transfor-mation $\sigma_m : \mathbb{R}^2 \to \mathbb{R}^2$ with the following properties:*

1. *If P is a point on m, then $\sigma_m(P) = P$, i.e., σ_m fixes line m pointwise.*

2. *If P is a point off m and $P' = \sigma_m(P)$, then m is the perpendicular bisector of $\overline{PP'}$.*

*The line m is called the **axis of reflection** (see Figure 4.6).*

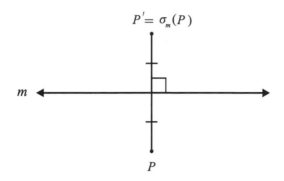

$P' = \sigma_m(P)$

m

P

Figure 4.6. Line m is the perpendicular bisector of $\overline{PP'}$.

Let us derive the equations of σ_m. Let m be a line with equation $aX + bY + c = 0$, with $a^2 + b^2 > 0$, and consider points $P = \begin{bmatrix} x \\ y \end{bmatrix}$ and $P' = \begin{bmatrix} x' \\ y' \end{bmatrix}$ such that $P' = \sigma_m(P)$. Assume for the moment that P is off m. By definition, $\overleftrightarrow{PP'} \perp m$. So if neither m nor $\overleftrightarrow{PP'}$ is vertical, the product of their respective slopes is

$$-\frac{a}{b} \cdot \frac{y' - y}{x' - x} = -1$$

or

$$\frac{y' - y}{x' - x} = \frac{b}{a}$$

and cross-multiplication gives

$$a(y' - y) = b(x' - x). \tag{4.2}$$

Note that Equation (4.2) also holds when m is vertical or horizontal. If m is vertical, its equation is $X + c = 0$, in which case $a = 1$ and $b = 0$. But reflection in a vertical line preserves the y-coordinate so that $y = y'$. On the other hand, if m is horizontal its equation is $Y + c = 0$, in which case $a = 0$ and $b = 1$. But reflection in a horizontal line preserves the x-coordinate so that $x = x'$. In either case, these are given by Equation (4.2). Now the midpoint M of P and P' has coordinates

$$M = \begin{bmatrix} \frac{x+x'}{2} \\ \frac{y+y'}{2} \end{bmatrix}.$$

Since M lies on m, its coordinates satisfy $aX + bY + c = 0$, which is the equation of line m. Therefore

$$a\left(\frac{x+x'}{2}\right) + b\left(\frac{y+y'}{2}\right) + c = 0. \tag{4.3}$$

Now rewrite Equations (4.2) and (4.3) to obtain the following system of linear equations in x' and y' :

$$\begin{cases} bx' - ay' = s \\ ax' + by' = t, \end{cases}$$

where $s = bx - ay$ and $t = -2c - ax - by$. Write this system in matrix form as

$$\begin{bmatrix} b & -a \\ a & b \end{bmatrix}\begin{bmatrix} x' \\ y' \end{bmatrix} = \begin{bmatrix} s \\ t \end{bmatrix}. \tag{4.4}$$

Since $a^2 + b^2 > 0$, the coefficient matrix is invertible and we may solve for x' and y' :

$$\begin{bmatrix} x' \\ y' \end{bmatrix} = \begin{bmatrix} b & -a \\ a & b \end{bmatrix}^{-1}\begin{bmatrix} s \\ t \end{bmatrix} = \frac{1}{a^2+b^2}\begin{bmatrix} b & a \\ -a & b \end{bmatrix}\begin{bmatrix} s \\ t \end{bmatrix} = \frac{1}{a^2+b^2}\begin{bmatrix} bs+at \\ bt-as \end{bmatrix}.$$

Substituting for s and t gives

$$\begin{aligned} x' &= \frac{1}{a^2+b^2}\left[b\left(bx - ay\right) + a\left(-2c - ax - by\right)\right] \\ &= \frac{1}{a^2+b^2}\left(b^2x - bay - 2ac - a^2x - aby\right) \\ &= \frac{1}{a^2+b^2}\left(b^2x + (a^2x - a^2x) - a^2x - 2aby - 2ac\right) \\ &= x - \frac{2a}{a^2+b^2}\left(ax + by + c\right), \end{aligned} \tag{4.5}$$

and similarly

$$y' = y - \frac{2b}{a^2+b^2}\left(ax + by + c\right). \tag{4.6}$$

Finally, if $P = \begin{bmatrix} x \\ y \end{bmatrix}$ is on m, then $ax + by + c = 0$ and Equations (4.5) and (4.6) reduce to

$$\begin{aligned} x' &= x \\ y' &= y, \end{aligned}$$

in which case P is a fixed point as required by the definition of σ_m. We have proved:

Theorem 128 *Let m be a line with equation $aX + bY + c = 0$, where $a^2 + b^2 > 0$. The equations of σ_m are:*

$$\begin{aligned} x' &= x - \frac{2a}{a^2+b^2}\left(ax + by + c\right) \\ y' &= y - \frac{2b}{a^2+b^2}\left(ax + by + c\right). \end{aligned} \tag{4.7}$$

Example 129 Let m be the line given by $X - Y + 5 = 0$. The equations for σ_m are:

$$
\begin{aligned}
x' &= x - (x - y + 5) &&= y - 5 \\
y' &= y - (-1)(x - y + 5) &&= x + 5.
\end{aligned}
$$

Thus $\sigma_m\left(\begin{bmatrix} 0 \\ 0 \end{bmatrix}\right) = \begin{bmatrix} -5 \\ 5 \end{bmatrix}$ and $\sigma_m\left(\begin{bmatrix} 5 \\ 5 \end{bmatrix}\right) = \begin{bmatrix} 0 \\ 10 \end{bmatrix}$.

The reflection of a vector $\mathbf{v} = \begin{bmatrix} x \\ y \end{bmatrix}$ in line $l : aX + bY + c = 0$ is the same for all c and depends only on the direction of l. In particular, thinking of $\mathbf{v}$ as a vector in standard position, its reflection in the line $m : aX + bY = 0$ is

$$
\sigma_m(\mathbf{v}) = \begin{bmatrix} x - \frac{2a}{a^2+b^2}(ax + by) \\ y - \frac{2b}{a^2+b^2}(ax + by) \end{bmatrix}. \tag{4.8}
$$

Then by straightforward algebra we can write Equation (4.8) in matrix form as

$$
\sigma_m(\mathbf{v}) = \frac{1}{a^2 + b^2} \begin{bmatrix} b^2 - a^2 & -2ab \\ -2ab & a^2 - b^2 \end{bmatrix} \mathbf{v}.
$$

Furthermore, if $\mathbf{u} = \begin{bmatrix} u_1 \\ u_2 \end{bmatrix}$ is a unit vector parallel to m, the (scalar) component of $\mathbf{v}$ parallel to $\mathbf{u}$ is the dot product $\mathbf{v} \cdot \mathbf{u} = xu_1 + yu_2$ and the projection of $\mathbf{v}$ on $\mathbf{u}$ is $(\mathbf{v} \cdot \mathbf{u})\mathbf{u}$. Note that $(\mathbf{v} \cdot \mathbf{u})\mathbf{u} - \mathbf{v}$ is orthogonal to $\mathbf{u}$ since $[(\mathbf{v} \cdot \mathbf{u})\mathbf{u} - \mathbf{v}] \cdot \mathbf{u} = (\mathbf{v} \cdot \mathbf{u})(\mathbf{u} \cdot \mathbf{u}) - \mathbf{v} \cdot \mathbf{u} = 0$, and $(\mathbf{v} \cdot \mathbf{u})\mathbf{u} = \mathbf{v} + [(\mathbf{v} \cdot \mathbf{u})\mathbf{u} - \mathbf{v}]$. Thus equation (4.8) can be expressed in vector form as

$$
\sigma_m(\mathbf{v}) = \mathbf{v} + 2[(\mathbf{v} \cdot \mathbf{u})\mathbf{u} - \mathbf{v}] = 2(\mathbf{v} \cdot \mathbf{u})\mathbf{u} - \mathbf{v}.
$$

Theorem 130 *Reflections are isometries.*

Proof. Let m be a line, let $\mathbf{u}$ be a unit vector parallel to m, and let P and Q be points. Let $P' = \sigma_m(P)$ and $Q' = \sigma_m(Q)$, and let $\mathbf{v} = \mathbf{PQ}$. Then

$$
P'Q' = \|\mathbf{P'Q'}\| = \|\sigma_m(\mathbf{PQ})\| = \|2(\mathbf{v} \cdot \mathbf{u})\mathbf{u} - \mathbf{v}\|, \text{ and}
$$

$$
\begin{aligned}
\|2(\mathbf{v} \cdot \mathbf{u})\mathbf{u} - \mathbf{v}\|^2 &= [2(\mathbf{v} \cdot \mathbf{u})\mathbf{u} - \mathbf{v}] \cdot [2(\mathbf{v} \cdot \mathbf{u})\mathbf{u} - \mathbf{v}] \\
&= 4(\mathbf{v} \cdot \mathbf{u})^2(\mathbf{u} \cdot \mathbf{u}) - 4(\mathbf{v} \cdot \mathbf{u})^2 + \mathbf{v} \cdot \mathbf{v} = \|\mathbf{v}\|^2 = \|\mathbf{PQ}\|^2.
\end{aligned}
$$

Taking square roots gives

$$
P'Q' = \|\mathbf{PQ}\| = PQ.
$$

Therefore the reflection σ_m is an isometry. ∎

We leave the proof of the following fact as an exercise for the reader:

Proposition 131 *Reflections are involutions.*

Since a reflection σ_m fixes its axis m pointwise, a non-trivial rotation $\rho_{C,\Theta}$ fixes only its center C, and a non-trivial translation is fixed point free, we have:

Corollary 132 *The set of translations, the set of non-trivial rotations, and the set of reflections are mutually disjoint sets.*

Trisecting a general angle with a straight edge and compass is a classical unsolvable problem. Interestingly, this problem has a solution when the straight edge and compass are replaced with a reflecting instrument such as a MIRA. Thus the trisection algorithm presented here is an important application of reflections.

Algorithm 133 (Angle Trisection) *Given arbitrary rays $\overrightarrow{OX}$ and $\overrightarrow{OY}$:*

1. *Choose a point P on $\overrightarrow{OX}$.*

2. *Locate the point S on $\overrightarrow{OX}$ such that $OP = PS$.*

3. *Construct lines m and n through P such that $m \perp \overrightarrow{OY}$ and $n \perp m$.*

4. *Locate the line l such that $\sigma_l(S)$ is on m and O is on $\sigma_l(n)$.*

5. *Let $R = \sigma_l(O)$.*

Then $m\angle SOR = 2m\angle ROY$ (see Figure 4.7).

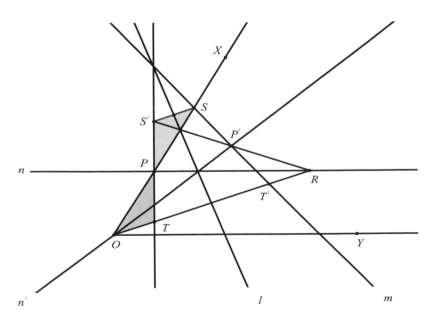

Figure 4.7. $m\angle SOR = 2m\angle ROY$.

Proof. Let $S' = \sigma_l(S)$ and $T = m \cap \overleftrightarrow{OR}$; then $\overleftrightarrow{SS'} \parallel \overleftrightarrow{OT}$ since $\overleftrightarrow{SS'}$ and $\overleftrightarrow{OT}$ are perpendicular to l, and $\angle TOP \cong \angle PSS'$ since these are alternate interior angles of parallels $\overleftrightarrow{SS'}$ and $\overleftrightarrow{OT}$ cut by transversal $\overleftrightarrow{OS}$. Furthermore, $\angle OPT \cong \angle SPS'$, since these angles are vertical, and $OP = PS$ by construction. Therefore $\triangle POT \cong \triangle PSS'$ (ASA) and $S'P = PT$ (CPCTC). Let $T' = \sigma_l(T)$, $P' = \sigma_l(P)$, and $n' = \sigma_l(n)$; then P' is on n' since P is on n, and $SP' = P'T'$ since reflections are isometries. Furthermore, $n' \perp \overleftrightarrow{ST'}$ since $n \perp \overleftrightarrow{S'T}$. Therefore n' is the perpendicular bisector of $\overline{ST'}$. Since O is on n' by construction, n' bisects $\angle SOT'$ so that $\angle SOP' \cong \angle P'OT' = \angle P'OR$. Now $\angle P'OR \cong \angle PRO$ since these angles are the reflections of each other in line l, and $\angle PRO \cong \angle ROY$ since these are alternate interior angles of parallels n and $\overleftrightarrow{OY}$ cut by transversal $\overleftrightarrow{OR}$. Therefore $\angle SOP' \cong \angle P'OR \cong \angle ROY$. ∎

For a theoretical discussion of MIRA constructions from the perspective of abstract algebra see [10].

Exercises

1. Words such as MOM and RADAR that spell the same forward and backward, are called *palindromes*.

    (a) When reflected in their vertical midlines, MOM remains MOM but the R's and D in RADAR appear backward. Find at least five other words like MOM that are preserved under reflection in their vertical midlines.

    (b) When reflected in their horizontal midlines, MOM becomes WOW, but BOB remains BOB. Find at least five other words like BOB that are preserved under reflection in their horizontal midlines.

2. Which capital letters could be cut out of paper and given a single fold to produce the figure below?

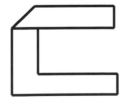

3. Given parallel lines p and q in the diagram below, use a MIRA to construct the path of a ray of light issuing from A and passing through B after being reflected exactly twice in p and once in q.

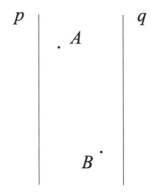

4. The diagram below shows a par 2 hole on a miniature golf course. Use a MIRA to construct the path the ball must follow to score a hole-in-one after banking the ball off

   (a) wall p.

   (b) wall q.

   (c) walls p and q.

   (d) walls p, q, and r.

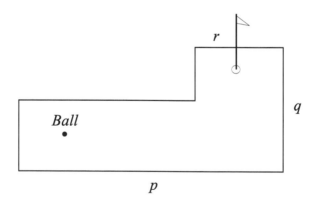

5. Two cities, located at points A and B in the diagram below, need to pipe water from the river, represented by line r. City A is 2 miles north of the river; city B is 10 miles downstream from A and 3 miles north of the river. The State will build one pumping station along the river.

   (a) Use a MIRA to locate the point C along the river at which the pumping station should be built so that the minimum amount of pipe is used to connect city A to C and city B to C.

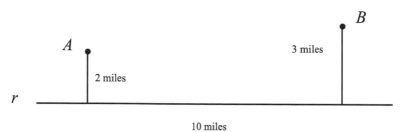

(b) Let A' be the reflection of A in r. Let $O = \overleftrightarrow{AA'} \cap r$ be the origin. Then $A = \begin{bmatrix} 0 \\ 2 \end{bmatrix}$ and $B = \begin{bmatrix} 10 \\ 3 \end{bmatrix}$. Compute the coordinates of the optimum point C constructed in part a.

(c) Compute the minimum $AC + CB$.

(d) Prove that if D is any point on r distinct from the optimum point C, then $AD + DB > AC + CB$.

6. Horizontal lines p and q in the diagram below have respective equations $Y = 0$ and $Y = 5$.

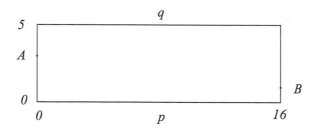

(a) Use a MIRA to construct the shortest path from point $A = \begin{bmatrix} 0 \\ 3 \end{bmatrix}$ to point $B = \begin{bmatrix} 16 \\ 1 \end{bmatrix}$ that first touches q and then p.

(b) Determine the coordinates of the point on q and the point on p touched by the path constructed in part a.

(c) Find the length of the path from A to B constructed in part a.

7. A river with parallel banks p and q is to be spanned by a bridge at right angles to p and q.

(a) Use a MIRA to locate the bridge that minimizes the distance from city A to city B.

(b) Let $\overline{PQ}$ denote the bridge at right angles to p and q constructed in part a. Prove that if $\overline{RS}$ is any other bridge spanning river r distinct from and parallel to $\overline{PQ}$, then $AR + RS + SB > AP + PQ + QB$.

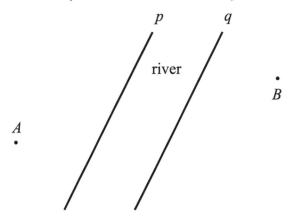

8. Suppose lines l and m intersect at the point Q, and let $l' = \sigma_m(l)$. Let P and R be points on l and l' that are distinct from Q and on the same side of m. Let S and T be the feet of the perpendiculars from P and R to m. Prove that $\angle PQS \cong \angle RQT$. (Thus when a ray of light is reflected by a flat mirror, the angle of incidence equals the angle of reflection.)

9. A ray of light is reflected by two perpendicular flat mirrors. Prove that the emerging ray is parallel to the initial incoming ray as indicated in the diagram below.

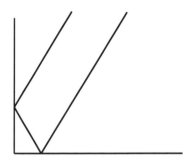

10. The Smiths, who range in height from 170 cm to 182 cm, wish to purchase a flat wall mirror that allows each Smith to view the full length of his or her image. Use the fact that each Smith's eyes are 10 cm below the top of his or her head to determine the minimum length of such a mirror.

11. Graph the line l with equation $X + 2Y - 6 = 0$ on graph paper. Plot the point $P = \begin{bmatrix} -5 \\ 3 \end{bmatrix}$ and use a MIRA to locate and mark its image P'. Visually estimate the coordinates $\begin{bmatrix} x' \\ y' \end{bmatrix}$ of the image point P' and record

your estimates. Write down the equations for the reflection σ_l and use them to compute the coordinates of the image point P'. Compare these analytical calculations with your visual estimates of the coordinates.

12. Fill in the missing entry in each row of the following table:

Equation of l	P	$\sigma_l(P)$	Equation of l	P	$\sigma_l(P)$
$X = 0$	$\begin{bmatrix} x \\ y \end{bmatrix}$	*	$Y = -3$	$\begin{bmatrix} x \\ y \end{bmatrix}$	*
$Y = 0$	*	$\begin{bmatrix} x \\ y \end{bmatrix}$	*	$\begin{bmatrix} 5 \\ 3 \end{bmatrix}$	$\begin{bmatrix} -8 \\ 3 \end{bmatrix}$
$Y = X$	$\begin{bmatrix} 3 \\ 2 \end{bmatrix}$	*	*	$\begin{bmatrix} 0 \\ 3 \end{bmatrix}$	$\begin{bmatrix} -3 \\ 0 \end{bmatrix}$
$Y = X$	$\begin{bmatrix} x \\ y \end{bmatrix}$	*	*	$\begin{bmatrix} -y \\ -x \end{bmatrix}$	$\begin{bmatrix} x \\ y \end{bmatrix}$
$X = 2$	*	$\begin{bmatrix} 6 \\ 3 \end{bmatrix}$	$Y = 2X$	$\begin{bmatrix} 0 \\ 5 \end{bmatrix}$	*
$Y = -3$	*	$\begin{bmatrix} -4 \\ -5 \end{bmatrix}$	$Y = X + 5$	$\begin{bmatrix} x \\ y \end{bmatrix}$	*

13. For each of the following pairs of points P and P', determine the equation of the axis l such that $P' = \sigma_l(P)$.

   (a) $P = \begin{bmatrix} 1 \\ 1 \end{bmatrix}$, $P' = \begin{bmatrix} -1 \\ -1 \end{bmatrix}$.

   (b) $P = \begin{bmatrix} 2 \\ 6 \end{bmatrix}$, $P' = \begin{bmatrix} 4 \\ 8 \end{bmatrix}$.

14. The equation of line l is $Y = 2X - 5$. Find the coordinates of the images of $\begin{bmatrix} 0 \\ 0 \end{bmatrix}$, $\begin{bmatrix} 1 \\ -3 \end{bmatrix}$, $\begin{bmatrix} -2 \\ 1 \end{bmatrix}$ and $\begin{bmatrix} 2 \\ 4 \end{bmatrix}$ under reflection in line l.

15. The equation of line m is $X - 2Y + 3 = 0$. Find the coordinates of the images of $\begin{bmatrix} 0 \\ 0 \end{bmatrix}$, $\begin{bmatrix} 4 \\ -1 \end{bmatrix}$, $\begin{bmatrix} -3 \\ 5 \end{bmatrix}$ and $\begin{bmatrix} 3 \\ 6 \end{bmatrix}$ under reflection in line m.

16. The equation of line p is $2X + 3Y + 4 = 0$; the equation of line q is $X - 2Y + 3 = 0$. Find the equation of the line $r = \sigma_q(p)$.

17. Let l and m be the lines with respective equations $X + Y - 2 = 0$ and $X + Y + 8 = 0$.

   (a) Compose the equations of σ_m and σ_l and show that the composition $\sigma_m \circ \sigma_l$ is a translation $\tau_{\mathbf{v}}$.

   (b) Compare $\|\mathbf{v}\|$ with the distance between l and m.

18. Let l and m be the lines with respective equations $X + Y - 2 = 0$ and $X - Y + 8 = 0$.

(a) Compose the equations of σ_m and σ_l and show that the composition $\sigma_m \circ \sigma_l$ is a halfturn φ_C.

(b) Determine the coordinates $\begin{bmatrix} a \\ b \end{bmatrix}$ of the center C of φ_C from the equations of φ_C derived in part a.

(c) Compute the coordinates of the point $P = l \cap m$ by solving the equations $X + Y - 2 = 0$ and $X - Y + 8 = 0$ simultaneously. Compare your result with your result in part b. What do you observe?

19. Let $P = \begin{bmatrix} p_1 \\ p_2 \end{bmatrix}$ and $Q = \begin{bmatrix} q_1 \\ q_2 \end{bmatrix}$ be distinct points and let $c : aX + bY + d = 0$ with $a^2 + b^2 > 0$ be a line parallel to $\overleftrightarrow{PQ}$. Prove that:

(a) $a(q_1 - p_1) + b(q_2 - p_2) = 0$.

(b) $\sigma_c \circ \tau_{\mathbf{PQ}} = \tau_{\mathbf{PQ}} \circ \sigma_c$.

20. Let l and m be lines such that $\sigma_m(l) = l$. Prove that either $l = m$ or $l \perp m$.

21. Find all values for a and b such that $\alpha\left(\begin{bmatrix} x \\ y \end{bmatrix}\right) = \begin{bmatrix} ay \\ x/b \end{bmatrix}$ is an involution.

22. Prove Proposition 131: Reflections are involutions.

23. Which reflections are linear? Explain.

4.4 Appendix: Geometer's Sketchpad Commands Required by Exploratory Activities

- To *select an object*, select the Translation Arrow Tool, then click on the object.

- To *select the line icon*, point to the small arrow in the lower right-hand corner of the Line Straightedge Tool, hold down your left mouse button and select the bidirectional arrow (the line symbol).

- To *construct a line*, select the line icon, click anywhere in the sketch, move your cursor a short distance in any direction and click again.

- To *label an object* with a default label, point to the object, right-click to display a menu, and select "Show Label."

- To *relabel an object*, display a dialog box by selecting the Translation Arrow Tool, then double clicking on the label displayed. Overwrite the label in the dialog box with the desired label and click ok.

- To *mark a line as a mirror*, select the line, then select [Mark Mirror] from the Transform menu.

- To *reflect an object in a mirror line m,* mark line m as a mirror line, select the object, then select [Reflect] from the Transform menu.

- To *construct a point*, select the Point Tool and click at the desired location.

- To *construct a segment*, select the endpoints, then select [Segment] from the Construct menu.

- To *construct a point of intersection,* select the intersecting objects, then select [Intersection] from the Construct menu.

- To *measure the distance between points X and Y,* select the points X and Y, then select [Distance] from the Measure menu.

- To *construct a point on a line*, select the line, then select [Point on Line] from the Construct menu.

- To *measure $\angle XYZ$,* select the points X, Y, and Z in that order, then select [Angle] from the Measure menu.

- To *mark vector* **XY**, select the points X and Y in that order, then select [Mark Vector] from the Construct menu.

- To *translate point Z by vector* **XY**, mark the vector **XY**, select the point X, then select [Translate] from the Transform menu.

- To *construct the n-gon with vertices* $V_1, V_2, \ldots, V_n$, select the points $V_1, V_2, \ldots, V_n$ in that order, then select [Segments] from the Construct menu.

- To *construct C_X,* select the points C and X in that order, then select [Circle by Center+Point] from the Construct menu.

- To *construct $\overrightarrow{XY}$,* select the points X and Y in that order, then select [Ray] from the Construct menu.

- To *mark a point as the center of rotation,* select the point, then select [Mark Center] from the Transform menu.

- To *rotate an object about point C through an angle of $\Theta°$,* mark the point C as the center of rotation, select the object, then select [Rotate] from the Transform menu, enter the rotation angle Θ, and click the Rotate button.

Chapter 5

Compositions of Translations, Rotations, and Reflections

Given a geometric figure and a rotation, reflection, or translation, draw the transformed figure using, e.g., graph paper, tracing paper, or geometry software. Specify a sequence of transformations that will carry a given figure onto another.

Common Core State Standards for Mathematics
CCSS.MATH.CONTENT.HSG.CO.A.5

According to Proposition 89, the composition of isometries is an isometry, so it is natural to study the properties possessed by a composition of isometries. In this chapter we shall observe that the composition of two reflections is either a rotation or a translation, the composition of rotations is a rotation or a translation, and the composition of a non-trivial rotation and translation is a non-trivial rotation. We conclude with a discussion of glide reflections, which are compositions of a translation and a reflection in an axis parallel to the direction of translation.

5.1 The Three Points Theorem

The goal of this section is to prove the Three Points Theorem, which asserts that an isometry is completely determined by its action on three non-collinear points. This powerful result will be applied in the sections that follow to characterize rotations and translations as a composition of two reflections. We begin with two preliminary results.

Theorem 134 *An isometry with distinct fixed points P and Q fixes the line $\overleftrightarrow{PQ}$ pointwise.*

Proof. Let P and Q be distinct points fixed by an isometry α, let R be any point on $\overleftrightarrow{PQ}$ distinct from P and Q, and let $R' = \alpha(R)$. Since α is an isometry, $PR = PR'$ and $QR = QR'$; hence $R' \in P_R \cap Q_R$ (see Figure 5.1). If $R' = R$

we're done, so assume $R' \neq R$. Then $\overleftrightarrow{PQ}$ is the perpendicular bisector of $\overline{RR'}$ since R and R' are equidistant from P and Q. Thus R is off $\overleftrightarrow{PQ}$, which is a contradiction. Therefore $R' = R$ for every point R on $\overleftrightarrow{PQ}$. ∎

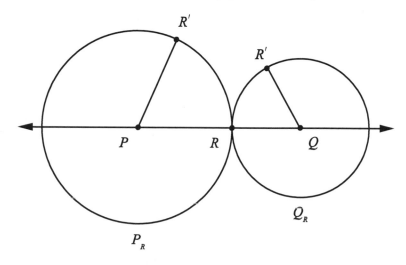

Figure 5.1. Proof of Theorem 134.

Theorem 135 *An isometry with three non-collinear fixed points is the identity.*

Proof. Let P, Q, and R be non-collinear points fixed by an isometry α. Then α fixes $\overleftrightarrow{PQ}$, $\overleftrightarrow{PR}$, and $\overleftrightarrow{QR}$ pointwise by Theorem 134. Let Z be any point off $\overleftrightarrow{PQ}$, $\overleftrightarrow{PR}$, and $\overleftrightarrow{QR}$, and let M be any point in the interior of $\triangle PQR$ distinct from Z (see Figure 5.2). Then $\overleftrightarrow{ZM}$ intersects $\triangle PQR$ in two distinct points A and B (one possibly a vertex). Since α fixes the points A and B, it fixes $\overleftrightarrow{ZM}$ pointwise by Theorem 134, and it follows that $\alpha = \iota$. ∎

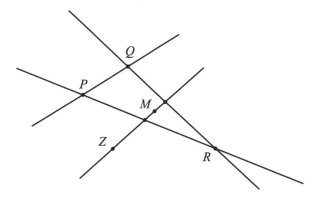

Figure 5.2. Proof of Theorem 135.

Theorem 136 (Three Points Theorem) *Two isometries that agree on three non-collinear points are equal.*

Proof. Suppose that isometries α and β agree on three non-collinear points P, Q, and R. Then

$$\alpha(P) = \beta(P), \quad \alpha(Q) = \beta(Q), \quad \text{and} \quad \alpha(R) = \beta(R). \tag{5.1}$$

Apply α^{-1} to both sides of each equation in (5.1) and obtain

$$P = (\alpha^{-1} \circ \beta)(P), \quad Q = (\alpha^{-1} \circ \beta)(Q), \quad \text{and} \quad R = (\alpha^{-1} \circ \beta)(R).$$

Then $\alpha^{-1} \circ \beta$ is an isometry that fixes three non-collinear points. Hence $\alpha^{-1} \circ \beta = \iota$ by Theorem 135, and an application of α to both sides gives $\alpha = \beta$. ∎

5.2   Rotations as Compositions of Two Reflections

We motivate the ideas in this section with an exploratory activity that requires the following definition:

Definition 137 *Given lines l and m, let $C \in l \cap m$, let A and B be points distinct from C on l and m, respectively, and let $A' = \varphi_C(A)$ and $B' = \varphi_C(B)$. Then $\angle ACB$, $\angle ACB'$, $\angle A'CB$, and $\angle A'CB'$ are the **angles from** l **to** m.*

In fact, *twice the measures of the angles from l to m are congruent* (mod 360). For example, the angles $\angle ACB$ and $\angle ACB'$ from l to m in Definition 137 are *supplementary* since they form a linear pair. Since the measures of the directed angles are signed, we have

$$m\angle ACB - m\angle ACB' = m\angle B'CA + m\angle ACB = m\angle B'CB = \pm 180$$

so that $2m\angle ACB - 2m\angle ACB' = \pm 360$ and $2m\angle ACB \equiv 2m\angle ACB'$.

Exploratory Activity 4: Rotations as compositions of two reflections.

1. Construct four distinct non-collinear points, label two points farthest apart C and D, and the remaining two points E and F.

2. Construct triangle $\triangle DEF$ and mark C as the center of rotation.

3. Rotate $\triangle DEF$ about C through an angle of $72°$ and label the corresponding vertices of the image triangle D', E', and F'.

4. Construct line $\overleftrightarrow{CD}$ and the perpendicular bisector of $\overline{DD'}$. Label these lines m and n. Where do m and n intersect? Explain.

5. Construct a point N on line n and measure the angle $\angle DCN$ from m to n. How is $2m\angle DCN$ related to the rotation angle $72°$?

6. Construct the point $N' = \varphi_C(N)$ and measure the angle $\angle DCN'$ from m to n. Note that $m\angle DCN$ and $m\angle DCN'$ have opposite sign. How is $2m\angle DCN'$ related to the rotation angle $72°$? How is $2m\angle DCN$ related to $2m\angle DCN'$?

7. Mark line m as a mirror, reflect $\triangle DEF$ in line m, and label the corresponding vertices of the image P, Q, and R.

8. Mark line n as a mirror and reflect $\triangle PQR$ in line n. Describe the image of $\triangle PQR$.

9. Complete the following sentence: *The rotation $\rho_{C,72}$ is a composition of reflections $\sigma_n \circ \sigma_m$, where the axes m and n* _____ *at C and twice the measure of an angle from m to n is congruent to* _____.

Theorem 138 *Given lines l and m, let $C \in l \cap m$ and let Θ be the measure of an angle from l to m. Then*

$$\sigma_m \circ \sigma_l = \rho_{C,2\Theta}.$$

Proof. If $l = m$, then $\Theta = 0$ and $\sigma_m \circ \sigma_l = \iota = \rho_{C,2\Theta}$. If $l \neq m$, first observe that

$$(\sigma_m \circ \sigma_l)(C) = \sigma_m(\sigma_l(C)) = \sigma_m(C) = C = \rho_{C,2\Theta}(C). \qquad (5.2)$$

Let L be a point on l distinct from C and consider the circle C_L. Let $M \in m \cap C_L$ such that $m\angle LCM = \Theta$ and let $L' = \sigma_m(L)$; then m is the perpendicular bisector of $\overline{LL'}$, by definition of σ_m, so that $CL = CL'$ and $m\angle LCL' = 2\Theta$ (see Figure 5.3). Therefore $L' = \rho_{C,2\Theta}(L)$ by definition of $\rho_{C,2\Theta}$ and

$$(\sigma_m \circ \sigma_l)(L) = \sigma_m(\sigma_l(L)) = \sigma_m(L) = L' = \rho_{C,2\Theta}(L). \qquad (5.3)$$

Let $M' = \sigma_l(M)$; then l is the perpendicular bisector of $\overline{MM'}$, by the definition of σ_l, so that $CM' = CM$ and $m\angle M'CM = 2\Theta$. Therefore $M = \rho_{C,2\Theta}(M')$ by definition of $\rho_{C,2\Theta}$ and

$$(\sigma_m \circ \sigma_l)(M') = \sigma_m(\sigma_l(M')) = \sigma_m(M) = M = \rho_{C,2\Theta}(M'). \qquad (5.4)$$

Hence the isometries $\sigma_m \circ \sigma_l$ and $\rho_{C,2\Theta}$ agree on non-collinear points C, L, and M' by Equations (5.2), (5.3), and (5.4). Therefore $\sigma_m \circ \sigma_l = \rho_{C,2\Theta}$ by Theorem 136. ∎

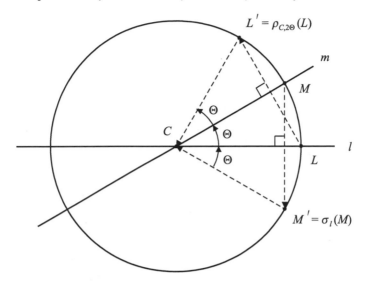

Figure 5.3. $\sigma_m \circ \sigma_l = \rho_{C,2\Theta}$.

In fact, *every* rotation is a composition of two reflections:

Theorem 139 *An isometry α is a rotation if and only if α is the composition of two reflections in lines that either intersect or are equal.*

Proof. ($\Leftarrow$) This implication is the statement in Theorem 138.
($\Rightarrow$) Consider a rotation $\rho_{C,2\Theta}$ and let $\Theta' \in (-180, 180]$ such that $\Theta' \equiv \Theta$. If $\Theta' = 0$ or 180, let l be any line and set $m = l$. Then $\rho_{C,2\Theta} = \iota = \sigma_m \circ \sigma_l$. If $\Theta' \neq 0$ or 180, let l be any line through C and let m be the unique line through C such that the measure of an angle from l to m is Θ' (see Figure 5.4). Then $\rho_{C,2\Theta} = \rho_{C,2\Theta'} = \sigma_m \circ \sigma_l$ by Theorem 138. ∎

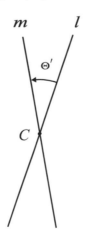

Figure 5.4. Proof of Theorem 139.

Example 140 Consider the lines $l : X - Y = 0$ and $m : X = 0$. The equations for reflections in l and m are

$$\sigma_l : \begin{cases} x' = y \\ y' = x \end{cases} \quad \text{and} \quad \sigma_m : \begin{cases} x' = -x \\ y' = y \end{cases}$$

and the equations for the composition $\sigma_m \circ \sigma_l$ are

$$\sigma_m \circ \sigma_l : \begin{cases} x' = -y \\ y' = x. \end{cases}$$

Note that the measure of the positive angle from l to m is 45, and the equations

$$\rho_{O,90} : \begin{cases} x' = x \cos 90 - y \sin 90 = -y \\ y' = x \sin 90 + y \cos 90 = x \end{cases}$$

agree with those of $\sigma_m \circ \sigma_l$. Furthermore, $-270 = 2(-135)$ is twice the measure of the negative angle from l to m, and of course, $90 \equiv -270$.

Theorem 141 *If lines l, m, and n are concurrent at C, then there exist unique lines p and q passing through C such that*

$$\sigma_m \circ \sigma_l = \sigma_n \circ \sigma_p = \sigma_q \circ \sigma_n.$$

Proof. If $l = m$, then $\sigma_m \circ \sigma_l = \iota = \sigma_n \circ \sigma_p = \sigma_q \circ \sigma_n$, where $p = q = n$. If $l \neq m$, let Θ be the measure of an angle from l to m. Let p be the unique line such that the measure of an angle from p to n is Θ. Then $\sigma_m \circ \sigma_l = \rho_{C,2\Theta} = \sigma_n \circ \sigma_p$ by Theorem 138. Similarly, let q be the unique line such that the measure of an angle from n to q is Θ. Then $\sigma_m \circ \sigma_l = \rho_{C,2\Theta} = \sigma_q \circ \sigma_n$ (see Figure 5.5). ∎

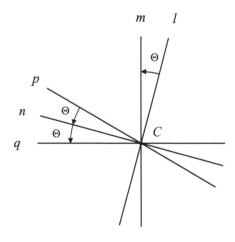

Figure 5.5. $\sigma_m \circ \sigma_l = \sigma_n \circ \sigma_p = \sigma_q \circ \sigma_n$.

Note that Theorem 141 does not require line n to be distinct from lines l and m. If $n = m$, for example, then $p = l$. Also, if $\rho_{C,\Theta} = \sigma_m \circ \sigma_l$, then $C \in l \cap m$. By Theorem 141, each line n passing through C determines unique lines p and q such that $\sigma_m \circ \sigma_l = \sigma_n \circ \sigma_p = \sigma_q \circ \sigma_n$. Since Θ is congruent to twice the measure of an angle from p to n and an angle from n to q we have:

Corollary 142 *Let n be a line passing through point C. Let p and q be the unique lines passing through C such that an angle from p to n and an angle from n to q is congruent to $\frac{1}{2}\Theta$. Then*

$$\rho_{C,\Theta} = \sigma_q \circ \sigma_n = \sigma_n \circ \sigma_p.$$

Corollary 143 *Let l and m be lines intersecting at point C. Then $l \perp m$ if and only if*

$$\varphi_C = \sigma_m \circ \sigma_l = \sigma_l \circ \sigma_m.$$

Proposition 123 tells us that a halfturn φ_C fixes a line l if and only if C is on l. Can lines be fixed by a general rotation?

Theorem 144 *A non-trivial rotation that fixes a line is a halfturn.*

Proof. Let l be a line and consider a non-trivial rotation $\rho_{C,\Theta}$ fixing l. Let m be the line through C perpendicular to l. By Corollary 142, there is a line n through C such that $\rho_{C,\Theta} = \sigma_n \circ \sigma_m$ (see Figure 5.6). Since $l \perp m$ we have

$$l = \rho_{C,\Theta}(l) = (\sigma_n \circ \sigma_m)(l) = \sigma_n(\sigma_m(l)) = \sigma_n(l)$$

so that σ_n fixes l, in which case $n = l$ or $n \perp l$, by Exercise 4.3.20. But if $n \perp l$, then l is perpendicular to both m and n, contradicting the fact that $\Theta \notin 0°$ and $m \cap n = C$. Therefore $n = l$ and $m \perp n$ so that $\rho_{C,2\Theta} = \sigma_n \circ \sigma_m = \varphi_C$ by Corollary 143. ∎

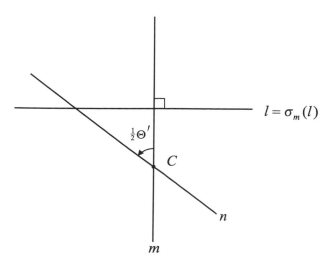

Figure 5.6. $\Theta' \equiv \Theta$.

Halfturns do not commute in general. In fact, $\varphi_A \circ \varphi_B = \varphi_B \circ \varphi_A$ if and only if $A = B$ (cf. Exercise 4.2.16). Thus *the only halfturn that commutes with* φ_A *is itself*.

Exercises

1. Consider the rotation $\rho_{C,\Theta} = \sigma_m \circ \sigma_l$, where $l : X = 3$ and $m : Y = X$.

   (a) Find the xy-coordinates of the center C and the rotation angle $\Theta' \in (-180, 180]$.

   (b) Find the equations of $\rho_{C,\Theta}$.

   (c) Compare Θ' with the (positive and negative) measures of the angles from l to m.

   (d) Compose the equations of σ_m with the equations of σ_l and compare your result with the equations of $\rho_{C,\Theta}$.

2. Consider the rotation $\rho_{C,\Theta} = \sigma_m \circ \sigma_l$, where $l : X + Y - 2 = 0$ and $m : Y = 3$.

   (a) Find the xy-coordinates of the center C and the rotation angle $\Theta' \in (-180, 180]$.

   (b) Find the equations of $\rho_{C,\Theta}$.

   (c) Compare Θ' with the (positive and negative) measures of the angles from l to m.

   (d) Compose the equations of σ_m with the equations of σ_l and compare your result with the equations of $\rho_{C,\Theta}$.

3. Consider the rotation $\rho_{C,\Theta} = \sigma_m \circ \sigma_l$, where $l : Y = X$ and $m : Y = -X + 4$.

   (a) Find the xy-coordinates of the center C and the rotation angle $\Theta' \in (-180, 180]$.

   (b) Find the equations of $\rho_{C,\Theta}$.

   (c) Compare Θ' with the (positive and negative) measures of the angles from l to m.

   (d) Compose the equations of σ_m with the equations of σ_l and compare your result with the equations of $\rho_{C,\Theta}$.

4. Find the equations of lines l and m such that $\rho_{O,90} = \sigma_m \circ \sigma_l$.

5. Let $C = \begin{bmatrix} 3 \\ 4 \end{bmatrix}$. Find equations of lines l and m such that $\rho_{C,60} = \sigma_m \circ \sigma_l$.

6. Consider the lines $l : X = 0$, $m : Y = 2X$, and $n : Y = 0$.

   (a) Find the equation of line p such that $\sigma_m \circ \sigma_l = \sigma_p \circ \sigma_n$.

   (b) Find the equation of line q such that $\sigma_m \circ \sigma_l = \sigma_n \circ \sigma_q$.

5.3 Translations as Compositions of Two Halfturns or Two Reflections

We motivate the ideas in this section with an exploratory activity.

Exploratory Activity 5: Translations as compositions of two reflections.

1. Construct four distinct non-collinear points, label two points farthest apart C and D, and the remaining two points E and F.

2. Construct triangle $\triangle DEF$ and mark the translation vector **CD**.

3. Translate $\triangle DEF$ by vector **CD** and label the corresponding vertices of the image triangle D', E', and F'.

4. Construct line $\overleftrightarrow{DD'}$, the line through D perpendicular to $\overleftrightarrow{DD'}$, and the perpendicular bisector of $\overline{DD'}$. Label these lines l, m, and n. How are the lines l, m, and n related?

5. Measure the distance CD and the distance from m to n. How do these distances compare?

6. Mark line m as a mirror, reflect $\triangle DEF$ in line m, and label the corresponding vertices of the image P, Q, and R.

7. Mark line n as a mirror and reflect $\triangle PQR$ in line n. Describe the image of $\triangle PQR$.

8. Complete the following sentence: *A translation τ_v is a composition of two reflections $\sigma_n \circ \sigma_m$, where the axes m and n are* _____ *to the direction of v and twice the distance from m to n is* _____ .

   Our next result is a special case of Theorem 157 (The Angle Addition Theorem, part 2) for halfturns, which asserts that the composition of two rotations whose rotation angle sum is a multiple of 360 is a translation. Corollary 146 follows immediately from Theorem 145 and identifies a composition of two reflections in parallel lines with a translation.

Theorem 145 *The composition of two halfturns is a translation. In fact, given any two points A and B,*

$$\varphi_B \circ \varphi_A = \tau_{2\mathbf{AB}}.$$

Proof. Given points $A = \begin{bmatrix} a \\ b \end{bmatrix}$ and $B = \begin{bmatrix} c \\ d \end{bmatrix}$, note that $2\mathbf{AB} = \begin{bmatrix} 2(c-a) \\ 2(d-b) \end{bmatrix}$. The equations of φ_A are $x' = 2a - x$ and $y' = 2b - y$, and the equations of φ_B are $x' = 2c - x$ and $y' = 2d - y$. Hence

$$(\varphi_B \circ \varphi_A)\left(\begin{bmatrix} x \\ y \end{bmatrix}\right) = \varphi_B\left(\begin{bmatrix} 2a - x \\ 2b - y \end{bmatrix}\right) = \begin{bmatrix} 2c - (2a - x) \\ 2d - (2b - y) \end{bmatrix}$$

$$= \begin{bmatrix} x + 2(c - a) \\ y + 2(d - b) \end{bmatrix} = \tau_{2\mathbf{AB}}\left(\begin{bmatrix} x \\ y \end{bmatrix}\right),$$

and it follows that $\varphi_B \circ \varphi_A = \tau_{2\mathbf{AB}}$. ∎

When A and B are distinct, Theorem 145 tells us that $\varphi_B \circ \varphi_A$ translates a distance $2AB$ in the direction from A to B (see Figure 5.7).

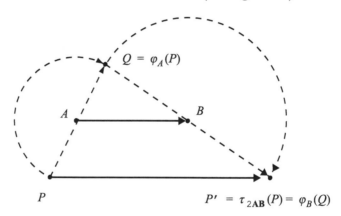

Figure 5.7. A composition of two halfturns.

Corollary 146 *Given parallel lines l and m, let $\mathbf{v}$ be the vector in the perpendicular direction from l to m whose magnitude is twice the distance from l to m. Then*

$$\sigma_m \circ \sigma_l = \tau_{\mathbf{v}}.$$

Proof. If $l = m$, then $\mathbf{v} = \mathbf{0}$ and $\sigma_m \circ \sigma_l = \iota = \tau_{\mathbf{v}}$. If $l \neq m$, let n be a common perpendicular, let $L = l \cap n$, and let $M = m \cap n$ (see Figure 5.8). Then $\mathbf{v} = 2\mathbf{LM}$ and by Corollary 143 and Theorem 145 we have

$$\sigma_m \circ \sigma_l = (\sigma_m \circ \sigma_n) \circ (\sigma_n \circ \sigma_l) = \varphi_M \circ \varphi_L = \tau_{2\mathbf{LM}} = \tau_{\mathbf{v}}. \qquad (5.5)$$

∎

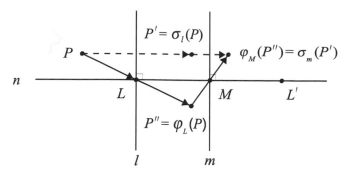

Figure 5.8. $\sigma_m \circ \sigma_l = \tau_{\mathbf{v}}$.

In fact, *every* translation is a composition two reflections:

Theorem 147 *An isometry α is a translation if and only if α is a composition of two reflections in parallel lines.*

Proof. ($\Leftarrow$) This implication is the statement in Corollary 146.

($\Rightarrow$) Given a translation τ, let L be a point and let $L' = \tau(L)$. Then $\mathbf{v} = \mathbf{LL'}$ and $\tau = \tau_{\mathbf{v}}$. If $L = L'$, choose any line l and set $m = l$. Then $\tau_{\mathbf{v}} = \iota = \sigma_m \circ \sigma_l$. If $L \neq L'$, let M be the midpoint of L and L', let $n = \overleftrightarrow{LM}$, and let l and m be the lines perpendicular to n at L and M, respectively. Then $\mathbf{v} = \mathbf{LL'} = 2\mathbf{LM}$, and $\tau_{\mathbf{v}} = \sigma_m \circ \sigma_l$ by Corollary 146. ∎

Thinking of a composition of two halfturns as a translation, our next result is a special case of Theorem 157 (The Angle Addition Theorem, parts 3 and 4) for halfturns, which asserts that the composition of a translation and a rotation of $\Theta°$ (in either order) is a rotation of $\Theta°$. Theorem 148 will be applied in our proof of Theorem 149, which is the analogue of Theorem 141 for translations.

Theorem 148 *The composition of three halfturns is a halfturn. In fact, given points A, B, and C, let D be the unique point such that $\mathbf{AB} = \mathbf{DC}$. Then*

$$\varphi_C \circ \varphi_B \circ \varphi_A = \varphi_D. \tag{5.6}$$

Proof. Let A, B, and C be three points, and let D be the unique point such that $\mathbf{AB} = \mathbf{DC}$. By Theorem 145, we have

$$\varphi_B \circ \varphi_A = \tau_{2\mathbf{AB}} = \tau_{2\mathbf{DC}} = \varphi_C \circ \varphi_D.$$

Since φ_C is an involution, applying φ_C on both sides, we have

$$\varphi_C \circ \varphi_B \circ \varphi_A = \varphi_C \circ \varphi_C \circ \varphi_D = \iota \circ \varphi_D = \varphi_D.$$

∎

Note that if A, B, and C are non-collinear, the center point D of the halfturn $\varphi_D = \varphi_C \circ \varphi_B \circ \varphi_A$ is the unique point such that $\square ABCD$ is a parallelogram (see Figure 5.9). This fact gives us a simple way to construct the center point D.

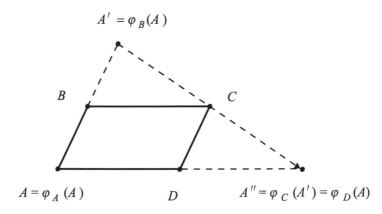

$$A' = \varphi_B(A)$$

$$B \qquad\qquad C$$

$$A = \varphi_A(A) \qquad D \qquad A'' = \varphi_C(A') = \varphi_D(A)$$

Figure 5.9. $\varphi_C \circ \varphi_B \circ \varphi_A = \varphi_D$.

Theorem 149 *If l, m, and n are parallel lines, then there exist unique lines p and q parallel to l such that*

$$\sigma_m \circ \sigma_l = \sigma_n \circ \sigma_p = \sigma_q \circ \sigma_n.$$

Proof. If $l = m$, then $\sigma_m \circ \sigma_l = \iota = \sigma_n \circ \sigma_p = \sigma_q \circ \sigma_n$, where $p = q = n$. If $l \neq m$, choose a common perpendicular c. Let $L = l \cap c$, $M = m \cap c$, and $N = n \cap c$. By Theorem 148, there exist unique points P and Q on c such that $\varphi_P = \varphi_N \circ \varphi_M \circ \varphi_L$ and $\varphi_Q = \varphi_M \circ \varphi_L \circ \varphi_N$. Composing φ_N with both sides of the first equation gives

$$\varphi_N \circ \varphi_P = \varphi_N \circ \varphi_N \circ \varphi_M \circ \varphi_L = \varphi_M \circ \varphi_L.$$

Composing both sides of the second equation with φ_N gives

$$\varphi_Q \circ \varphi_N = \varphi_M \circ \varphi_L \circ \varphi_N \circ \varphi_N = \varphi_M \circ \varphi_L.$$

Then by Theorem 145, we have

$$\tau_{2LM} = \varphi_M \circ \varphi_L = \varphi_N \circ \varphi_P = \tau_{2PN}$$
$$= \varphi_M \circ \varphi_L = \varphi_Q \circ \varphi_N = \tau_{2NQ}.$$

Let p and q be the lines perpendicular to c at P and Q, respectively (see Figure 5.10). Then by Corollary 146, we have

$$\sigma_m \circ \sigma_l = \tau_{2LM} = \tau_{2PN} = \sigma_n \circ \sigma_p$$
$$= \tau_{2LM} = \tau_{2NQ} = \sigma_q \circ \sigma_n.$$

■

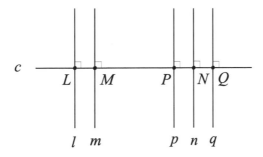

Figure 5.10. $\sigma_m \circ \sigma_l = \sigma_n \circ \sigma_p = \sigma_q \circ \sigma_n$.

Note that Theorem 149 does not require line n to be distinct from l and m. If $n = m$, for example, then $p = l$. Now if $\tau_{\mathbf{PQ}} = \sigma_m \circ \sigma_l$, then $l \parallel m$. If n is also parallel to m, Theorem 149 tells us that n determines unique lines p and q parallel to m such that $\sigma_m \circ \sigma_l = \sigma_n \circ \sigma_p = \sigma_q \circ \sigma_n$. Since $\mathbf{PQ} = 2\mathbf{PN} = 2\mathbf{NQ}$ we have:

Corollary 150 *Let P and Q be distinct points and let n be a line perpendicular to $\overleftrightarrow{PQ}$. Then there exist unique lines p and q parallel to n such that*

$$\tau_{\mathbf{PQ}} = \sigma_q \circ \sigma_n = \sigma_n \circ \sigma_p.$$

We collect our results in the following theorem:

Theorem 151 *A composition of two reflections is a translation or a rotation. Only the identity is both a translation and a rotation.*

Theorem 151 raises an interesting and important question: What is the result of composing *more than two* reflections? We shall give a complete answer to this question in Chapter 6.

Exercises

1. Lines l and m have respective equations $Y = 3$ and $Y = 5$. Find the equations of the translation $\sigma_m \circ \sigma_l$.

2. Lines l and m have respective equations $Y = X$ and $Y = X + 4$. Find the equations of the translation $\sigma_m \circ \sigma_l$.

3. The translation τ has vector $\begin{bmatrix} 4 \\ -3 \end{bmatrix}$. Find the equations of lines l and m such that $\tau = \sigma_m \circ \sigma_l$.

4. The translation τ has equations $x' = x + 6$ and $y' = y - 3$. Find equations of lines l and m such that $\tau = \sigma_m \circ \sigma_l$.

5. Consider the lines $l : Y = 3$, $m : Y = 5$, and $n : Y = 9$.

    (a) Find the equation of line p such that $\sigma_m \circ \sigma_l = \sigma_p \circ \sigma_n$.

    (b) Find the equation of line q such that $\sigma_m \circ \sigma_l = \sigma_n \circ \sigma_q$.

6. In the figure below, construct:

    (a) Line s such that $\sigma_n \circ \sigma_s = \sigma_m \circ \sigma_l$.

    (b) Line t such that $\sigma_c \circ \sigma_t = \sigma_b \circ \sigma_a$.

    (c) The fixed point of $\sigma_t \circ \sigma_s$.

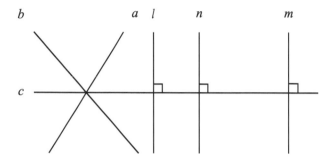

7. If coin A in the figure below is rolled around coin B until coin A is directly under coin B, will the head on coin A be right-side up or upside down? Explain.

(A)

(B)

8. In the figure below, sketch points X, Y, Z such that

    (a) $\varphi_A \circ \varphi_E \circ \varphi_D = \varphi_X$.

(b) $\varphi_D \circ \tau_{AC} = \varphi_Y$.

(c) $\tau_{BC} \circ \tau_{AB} \circ \tau_{EA}(A) = Z$.

9. Let A, B, and P be distinct points and let $P' = \tau_{AB}(P)$. Prove that $\tau_{AB} \circ \varphi_P \circ \tau_{BA} = \varphi_{P'}$.

10. Let τ be a translation, let C be a point, and let D be the midpoint of C and $\tau(C)$.

   (a) Prove that $\tau \circ \varphi_C = \varphi_D$.

   (b) Prove that $\varphi_C \circ \tau = \varphi_E$, where $E = \tau^{-1}(D)$.

11. Given three points A, B, and C, construct the point D such that $\tau_{AB} = \varphi_D \circ \varphi_C$.

12. Given three points A, B, and D, construct the point C such that $\tau_{AB} = \varphi_D \circ \varphi_C$.

13. Given three points A, C, and D, construct the point B such that $\tau_{AB} = \varphi_D \circ \varphi_C$.

14. Given three points B, C, and D, construct the point A such that $\tau_{AB} = \varphi_D \circ \varphi_C$.

15. Given any three points A, B and C, prove that $\varphi_C \circ \varphi_B \circ \varphi_A = \varphi_A \circ \varphi_B \circ \varphi_C$.

16. Given a point A and a non-trivial translation $\tau_{\mathbf{v}}$, construct the point B such that $\varphi_A \circ \tau_{\mathbf{v}} = \varphi_B$.

17. Given a point A and a non-trivial translation $\tau_{\mathbf{v}}$, construct the point B such that $\tau_{\mathbf{v}} \circ \varphi_A = \varphi_B$.

18. Let P be a point and let a and b be lines. Prove that:

    (a) There exist lines c and d with P on c such that $\sigma_b \circ \sigma_a = \sigma_d \circ \sigma_c$.

    (b) There exist lines l and m with P on m such that $\sigma_b \circ \sigma_a = \sigma_m \circ \sigma_l$.

19. Let P and Q be distinct points and let c be a line parallel to $\overleftrightarrow{PQ}$. Apply Corollaries 146 and 143 to prove that $\sigma_c \circ \tau_{\mathbf{PQ}} = \tau_{\mathbf{PQ}} \circ \sigma_c$ (cf. Exercise 4.3.19).

20. Let C be a point on line l and let $\Theta \in \mathbb{R}$. Prove that $\sigma_l \circ \rho_{C,\Theta} = \rho_{C,-\Theta} \circ \sigma_l$.

21. Prove that in any non-degenerate triangle, the perpendicular bisectors of the sides are concurrent at some point D equidistant from the vertices. Thus every triangle has a circumscribed circle, called the *circumcircle*. The center D of the circumcircle is called the *circumcenter* of the triangle.

5.4   The Angle Addition Theorem

The five statements in Theorem 157 (The Angle Addition Theorem) completely determine all compositions of translations and rotations. We motivate part 1 of this theorem with an exploratory activity.

Exploratory Activity 6: Composing rotations whose rotation angle sum is not a multiple of 360.

1. Construct distinct points A and B and line segment $\overline{AB}$.

2. Mark center of rotation B then rotate A through angles of $90°$ and $180°$ about B; label the image points C and D, respectively.

3. Mark center of rotation C, rotate $\overline{AB}$ $60°$ about C; label the image endpoints A' and B' with A' corresponding to A and B' corresponding to B.

4. Mark center of rotation D, rotate $\overline{A'B'}$ $40°$ about D; label the image endpoints A'' and B'' with A'' corresponding to A' and B'' corresponding to B'.

5. Construct segments $\overline{AA''}$ and $\overline{BB''}$ and their perpendicular bisectors. Label the intersection of perpendicular bisectors E.

6. Hide segments $\overline{AA''}$, $\overline{BB''}$, their midpoints and perpendicular bisectors.

7. Mark center of rotation E and rotate $\overline{AB}$ 20° about E.

8. Rotate the image of $\overline{AB}$ in step (7) 20° about E.

9. Rotate the image in step (8) 20° about E, and continue rotating in this manner two more iterations until the total rotation angle about E is 100°. What do you observe?

10. Complete the following conjecture: *The rotation $\rho_{C,60}$ followed by the rotation $\rho_{D,40}$ is the rotation* _____.

Using a technique similar to the one we used in the proof of Theorem 149, our proof of The Angle Addition Theorem transforms a composition of two rotations or a composition of a rotation and a translation into a composition of two reflections, which is either a rotation or a translation.

First, we consider a composition of two rotations whose rotation angle sum is not a multiple of 360.

Theorem 152 (The Angle Addition Theorem, part 1) *Let A and B be points and let Θ and Φ be real numbers such that $\Theta + \Phi \notin 0°$. There is a unique point C such that*

$$\rho_{B,\Phi} \circ \rho_{A,\Theta} = \rho_{C,\Theta+\Phi}.$$

Proof. If $A = B$, then $\rho_{B,\Phi} \circ \rho_{A,\Theta} = \rho_{B,\Phi} \circ \rho_{B,\Theta} = \rho_{B,\Theta+\Phi}$ by Proposition 116, and the conclusion holds with $C = B$. So assume that $A \neq B$ and let $\Theta', \Phi' \in (-180, 180]$ such that $\Theta' \equiv \Theta$ and $\Phi' \equiv \Phi$; then $\rho_{A,\Theta} = \rho_{A,\Theta'}$ and $\rho_{B,\Phi} = \rho_{B,\Phi'}$. If $\Theta' = 0$, then $\rho_{A,\Theta'} = \iota$ and the conclusion holds for $C = B$; similarly, if $\Phi' = 0$, the conclusion holds with $C = A$. So assume $\Theta', \Phi' \neq 0$ and let $m = \overrightarrow{AB}$. By Corollary 142, there exist unique lines l and n passing through A and B, respectively, such that $\rho_{B,\Phi'} = \sigma_n \circ \sigma_m$ and $\rho_{A,\Theta'} = \sigma_m \circ \sigma_l$. Consider the angle from l to m measuring $\frac{1}{2}\Theta'$ and the angle from m to n measuring $\frac{1}{2}\Phi'$. The assumption that $-360 < \Theta' + \Phi' < 360$ implies that $l \nparallel n$, for otherwise m is a transversal and congruent corresponding angles give $\frac{1}{2}\Theta' + \frac{1}{2}\Phi' = 180$ so that $\Theta' + \Phi' = 360$, which is a contradiction. Therefore l and n intersect at some point C and we may consider $\triangle ABC$. Now $\Theta', \Phi' \in (-180, 0) \cup (0, 180]$ implies $\frac{1}{2}\Theta', \frac{1}{2}\Phi' \in (-90, 0) \cup (0, 90]$. We claim that some angle from l to n has measure $\frac{1}{2}(\Theta' + \Phi')$. We consider four cases.

<u>Case 1:</u> $\frac{1}{2}\Theta', \frac{1}{2}\Phi' \in (0, 90]$. Then $m\angle CAB = \frac{1}{2}\Theta'$ and $m\angle ABC = \frac{1}{2}\Phi'$ (see Figure 5.11). By the Exterior Angle Theorem, there is an exterior angle from l to n with measure $\frac{1}{2}(\Theta' + \Phi')$.

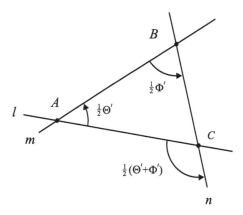

Figure 5.11. Proof of Theorem 152, Case 1.

<u>Case 2:</u> $\frac{1}{2}\Theta' \in (-90, 0)$ and $\frac{1}{2}\Phi' \in (0, 90]$; then $m\angle BAC = -\frac{1}{2}\Theta'$ and the interior angle $\angle ACB$ is an angle from l to n (see Figure 5.12). By the Exterior Angle Theorem, $\frac{1}{2}\Phi' = m\angle ACB - \frac{1}{2}\Theta'$; hence $m\angle ACB = \frac{1}{2}(\Theta' + \Phi')$.

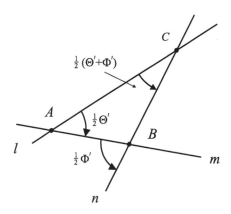

Figure 5.12. Proof of Theorem 152, Case 2.

Cases 3 and 4 are similar and left as exercises. Thus in every case, $\rho_{B,\Phi} \circ \rho_{A,\Theta} = \rho_{B,\Phi'} \circ \rho_{A,\Theta'} = (\sigma_n \circ \sigma_m) \circ (\sigma_m \circ \sigma_l) = \sigma_n \circ \sigma_l = \rho_{C,\Theta'+\Phi'} = \rho_{C,\Theta+\Phi}$. ∎

Example 153 Let $A = \begin{bmatrix} 2 \\ 2 \end{bmatrix}$ and $B = \begin{bmatrix} -2 \\ 2 \end{bmatrix}$. To determine $\rho_{B,180} \circ \rho_{A,90}$, let $m = \overleftrightarrow{AB} : Y = 2$, $l : Y = -X + 4$, and $n : X = -2$. Then l and n are the unique lines such that an angle from l to m measures 45 and an angle from m to n measures 90. Thus $\rho_{A,90} = \sigma_m \circ \sigma_l$, $\rho_{B,180} = \sigma_n \circ \sigma_m$, the center of rotation is $C = l \cap m = \begin{bmatrix} -2 \\ 6 \end{bmatrix}$, and

$$\rho_{B,180} \circ \rho_{A,90} = (\sigma_n \circ \sigma_m) \circ (\sigma_m \circ \sigma_l) = \sigma_n \circ \sigma_l = \rho_{C,-90} = \rho_{C,270}.$$

Second, we consider a composition of two rotations whose rotation angle sum is a multiple of 360.

Theorem 154 (The Angle Addition Theorem, part 2) *Let A and B be points and let Θ and Φ be real numbers such that $\Theta + \Phi \in 0°$. Then $\rho_{B,\Phi} \circ \rho_{A,\Theta}$ is a translation.*

Proof. If $A = B$, then $\rho_{B,\Phi} \circ \rho_{A,\Theta} = \rho_{B,\Theta+\Phi} = \rho_{B,0} = \iota = \tau_0$ (the trivial translation), by Proposition 116. So assume $A \neq B$ and let $m = \overleftrightarrow{AB}$. Then by Corollary 142, there exist unique lines l and n passing through A and B, respectively, such that $\rho_{B,\Phi} = \sigma_n \circ \sigma_m$ and $\rho_{A,\Theta} = \sigma_m \circ \sigma_l$. Since $\Theta + \Phi \in 0°$, either both Θ and Φ are multiples of 360 or neither Θ nor Φ is a multiple of 360. If $\Theta, \Phi \in 0°$, then $\rho_{A,\Theta} = \rho_{B,\Phi} = \iota$ so that $\rho_{B,\Phi} \circ \rho_{A,\Theta} = \iota = \tau_0$ (the trivial translation). If $\Theta, \Phi \notin 0°$, let $\Theta' = \Theta° \cap (0, 360)$ and $\Phi' = \Phi° \cap (0, 360)$. Then l, m, and n are distinct, some angle from l to m measures $\frac{1}{2}\Theta'$, some angle from m to n measures $\frac{1}{2}\Phi'$, and $\frac{1}{2}\Theta' + \frac{1}{2}\Phi' = 180$. Thus m cuts l and n with congruent corresponding angles and $l \parallel n$ (see Figure 5.13). Therefore

$$\rho_{B,\Phi} \circ \rho_{A,\Theta} = \rho_{B,\Phi'} \circ \rho_{A,\Theta'} = (\sigma_n \circ \sigma_m) \circ (\sigma_m \circ \sigma_l) = \sigma_n \circ \sigma_l,$$

which is a non-trivial translation by Theorem 147. ∎

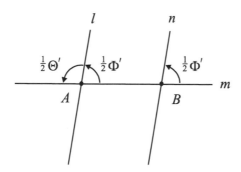

Figure 5.13. The translation $\rho_{B,\Phi} \circ \rho_{A,\Theta}$.

Example 155 Let $A = \begin{bmatrix} 2 \\ 2 \end{bmatrix}$ and $B = \begin{bmatrix} -2 \\ 2 \end{bmatrix}$. To determine $\rho_{B,90} \circ \rho_{A,270}$, let $m = \overleftrightarrow{AB} : Y = 2$, $l : Y = X$, and $n : Y = X + 4$. Then l and n are the unique lines such that an angle from l to m measures 135 and an angle from m to n measures 45. Thus $l \parallel n$, $\rho_{B,90} = \sigma_n \circ \sigma_m$, $\rho_{A,270} = \sigma_m \circ \sigma_l$, and the vector in the perpendicular direction from l to n is **OB**. Therefore

$$\rho_{B,180} \circ \rho_{A,90} = (\sigma_n \circ \sigma_m) \circ (\sigma_m \circ \sigma_l) = \sigma_n \circ \sigma_l = \tau_{2\mathbf{OB}}.$$

Third, we consider the composition of a translation and a non-trivial rotation (in either order).

Theorem 156 (The Angle Addition Theorem, parts 3 and 4)   *The composition of a non-trivial rotation of $\Theta°$ and a translation (in either order) is a rotation of $\Theta°$.*

Proof. Let $\rho_{C,\Theta}$ be a rotation such that $\Theta \notin 0°$. If $\tau = \iota$ there is nothing to prove. So assume $\tau \neq \iota$, and let m be the line through C perpendicular to the direction of translation. Let l and n be the unique lines such that $\tau = \sigma_n \circ \sigma_m$ and $\rho_{C,\Theta} = \sigma_m \circ \sigma_l$. Since $\rho_{C,\Theta} \neq \iota$, lines l and m are distinct and intersect at C. Let $\Theta' = \Theta° \cap (-180, 180]$. Since $\tau \neq \iota$, lines m and n are distinct and parallel. Hence l is a transversal for parallels m and n, and the corresponding angles from l to m and from l to n have measure $\frac{1}{2}\Theta'$ (see Figure 5.14). Let $D = l \cap n$; then

$$\tau \circ \rho_{C,\Theta} = \sigma_n \circ \sigma_m \circ \sigma_m \circ \sigma_l = \sigma_n \circ \sigma_l = \rho_{D,\Theta}.$$

The composition $\rho_{C,\Theta} \circ \tau$ is also a rotation of $\Theta°$ by a similar argument left to the reader. ∎

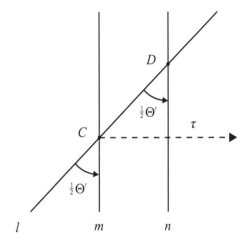

Figure 5.14. $\tau \circ \rho_{C,\Theta} = \rho_{D,\Theta}$.

Finally, we gather the various parts of the Angle Addition Theorem proved in this section together with Proposition 106, part 1.

Theorem 157 (The Angle Addition Theorem) *Let $\Theta, \Phi \in \mathbb{R}$.*

1. *If $\Theta + \Phi \notin 0°$, a rotation of $\Theta°$ followed by a rotation of $\Phi°$ is a rotation of $(\Theta + \Phi)°$.*

2. *If $\Theta + \Phi \in 0°$, a rotation of $\Theta°$ followed by a rotation of $\Phi°$ is a translation.*

3. *A non-trivial rotation of $\Theta°$ followed by a translation is a rotation of $\Theta°$.*

4. A translation followed by a non-trivial rotation of $\Theta°$ is a rotation of $\Theta°$.

5. A translation followed by a translation is a translation.

Exercises

1. Let $O = \begin{bmatrix} 0 \\ 0 \end{bmatrix}$ and $C = \begin{bmatrix} 2 \\ 0 \end{bmatrix}$.

   (a) Find equations of lines l, m, and n such that $\rho_{C,90} = \sigma_m \circ \sigma_n$ and $\rho_{O,90} = \sigma_l \circ \sigma_m$.

   (b) Find xy-coordinates of the point D such that $\varphi_D = \rho_{O,90} \circ \rho_{C,90}$.

   (c) Find xy-coordinates for the point E such that $\varphi_E = \rho_{C,90} \circ \rho_{O,90}$.

2. Let $O = \begin{bmatrix} 0 \\ 0 \end{bmatrix}$ and $C = \begin{bmatrix} 0 \\ 1 \end{bmatrix}$.

   (a) Find equations of the lines l, m, and n such that $\varphi_O = \sigma_m \circ \sigma_l$ and $\rho_{C,120} = \sigma_n \circ \sigma_m$.

   (b) Find xy coordinates for the point D and the angle of rotation Θ such that $\rho_{D,\Theta} = \rho_{C,120} \circ \varphi_O$.

   (c) Find xy-coordinates for the point E and the angle of rotation Φ such that $\rho_{E,\Phi} = \rho_{C,120} \circ \rho_{O,60}$.

3. Let $A = \begin{bmatrix} 4 \\ 0 \end{bmatrix}$ and $B = \begin{bmatrix} 0 \\ 4 \end{bmatrix}$.

   (a) Find equations of lines l, m, and n such that $\rho_{A,90} = \sigma_m \circ \sigma_l$ and $\rho_{B,120} = \sigma_n \circ \sigma_m$.

   (b) Find xy coordinates for the point C and the angle of rotation Θ such that $\rho_{C,\Theta} = \rho_{B,120} \circ \rho_{A,90}$.

   (c) Find xy-coordinates for the point D and the angle of rotation Φ such that $\rho_{D,\Phi} = \rho_{A,90} \circ \rho_{B,120}$.

4. Given distinct points A and B, let $\rho_{C,150} = \rho_{B,90} \circ \rho_{A,60}$ and $m = \overleftrightarrow{AB}$. Use a MIRA to construct lines l and n such that $\rho_{A,60} = \sigma_m \circ \sigma_l$ and $\rho_{B,90} = \sigma_n \circ \sigma_m$. Label the center of rotation C.

5. Given distinct points A and B, let $\rho_{C,30} = \rho_{B,90} \circ \rho_{A,-60}$ and $m = \overleftrightarrow{AB}$. Use a MIRA to construct lines l and n such that $\rho_{A,-60} = \sigma_m \circ \sigma_l$ and $\rho_{B,90} = \sigma_n \circ \sigma_m$. Label the center of rotation C.

6. Given distinct points A and B, let $\rho_{C,-30} = \rho_{B,-90} \circ \rho_{A,60}$ and $m = \overleftrightarrow{AB}$. Use a MIRA to construct lines l and n such that $\rho_{A,60} = \sigma_m \circ \sigma_l$ and $\rho_{B,-90} = \sigma_n \circ \sigma_m$. Label the center of rotation C.

7. Given distinct points A and B, let $\rho_{C,-150} = \rho_{B,-90} \circ \rho_{A,-60}$ and $m = \overleftrightarrow{AB}$. Use a MIRA to construct lines l and n such that $\rho_{A,-60} = \sigma_m \circ \sigma_l$ and $\rho_{B,-90} = \sigma_n \circ \sigma_m$. Label the center of rotation C.

8. Given distinct points A and B, let $\tau_{\mathbf{v}} = \rho_{B,120} \circ \rho_{A,240}$. Use a MIRA to construct lines l and n such that $\rho_{A,240} = \sigma_m \circ \sigma_l$ and $\rho_{B,120} = \sigma_n \circ \sigma_m$. Construct the translation vector $\mathbf{v}$.

9. Given a point A and a non-trivial translation $\tau_{\mathbf{v}}$, let $\rho_{B,90} = \tau_{\mathbf{v}} \circ \rho_{A,90}$. Use a MIRA to construct lines l, m, and n such that $\rho_{A,90} = \sigma_m \circ \sigma_l$ and $\tau_{\mathbf{v}} = \sigma_n \circ \sigma_m$. Label the center of rotation B.

10. Given a point A and a non-trivial translation $\tau_{\mathbf{v}}$, let $\rho_{B,90} = \rho_{A,90} \circ \tau_{\mathbf{v}}$. Use a MIRA to construct lines l, m, and n such that $\tau_{\mathbf{v}} = \sigma_m \circ \sigma_l$ and $\rho_{A,90} = \sigma_n \circ \sigma_m$. Label the center of rotation B.

11. Given distinct points P and Q, use a MIRA to construct the point R such that $\tau_{\mathbf{PQ}} \circ \rho_{P,45} = \rho_{R,45}$.

12. Given distinct non-collinear points P, Q, and R, use a MIRA to construct the point S such that $\tau_{\mathbf{QR}} \circ \rho_{P,120} = \rho_{S,120}$.

13. Given distinct points B and C, use a MIRA to construct the point A such that $\tau_{\mathbf{AB}} \circ \rho_{B,60} = \rho_{C,60}$.

14. Given distinct points B and C, use a MIRA to construct the point A such that $\rho_{B,60} \circ \tau_{\mathbf{AB}} = \rho_{C,60}$.

15. Given a pair of congruent rectangles $\square ABCD \cong \square EFGH$, construct a point Q and an angle of measure Θ such that $\rho_{Q,\Theta}(\square ABCD) = \square EFGH$.

16. Complete the proof of Theorem 152: Let A and B be points, and let Θ and Φ be real numbers such that $\Theta + \Phi \notin 0°$.

   (a) Case 3: Assume $\frac{1}{2}\Theta', \frac{1}{2}\Phi' \in (-90, 0)$ and prove that $\rho_{B,\Phi} \circ \rho_{A,\Theta} = \rho_{C,\Theta+\Phi}$.

   (b) Case 4: Assume $\frac{1}{2}\Theta' \in (0, 90]$ and $\Phi' \in (-90, 0)$ and prove that $\rho_{B,\Phi} \circ \rho_{A,\Theta} = \rho_{C,\Theta+\Phi}$.

17. Complete the proof of Theorem 156: Let C be a point, let $\Theta \notin 0°$, and let τ be a translation. Prove there exists a point D such that $\rho_{C,\Theta} \circ \tau = \rho_{D,\Theta}$.

5.5   Glide Reflections

Consider three lines l, m, and n (not necessarily distinct), which are either concurrent or mutually parallel. If they are concurrent at some point C, Theorem 141 asserts that there exists a unique line p passing through C such that $\sigma_m \circ \sigma_l = \sigma_n \circ \sigma_p$. Similarly, if they are mutually parallel, Theorem 149 asserts that there exists a unique line p parallel to l, m, and n such that $\sigma_m \circ \sigma_l = \sigma_n \circ \sigma_p$. In both cases, composing σ_n with both sides of the equation $\sigma_m \circ \sigma_l = \sigma_n \circ \sigma_p$ gives $\sigma_n \circ \sigma_m \circ \sigma_l = \sigma_n \circ \sigma_n \circ \sigma_p = \sigma_p$ and the following corollary:

Corollary 158 *Let l, m, and n be lines.*

   a. *If l, m, and n are concurrent at point C, there exists a unique line p passing through C such that*

$$\sigma_n \circ \sigma_m \circ \sigma_l = \sigma_p.$$

   b. *If l, m, and n are mutually parallel, there exists a unique line p parallel to l, m, and n such that*

$$\sigma_n \circ \sigma_m \circ \sigma_l = \sigma_p.$$

Thus the composition of three reflections in concurrent or mutually parallel lines is a reflection in some unique line concurrent with or mutually parallel to them. The converse is also true (see Exercise 5.5.7).

Proposition 159 *Lines l, m, and n are either concurrent or mutually parallel if and only if there exists a unique line p such that $\sigma_n \circ \sigma_m \circ \sigma_l = \sigma_p$.*

But what results from composing three reflections in three lines l, m, and n that are neither concurrent nor mutually parallel? Proposition 159 tells us that $\sigma_n \circ \sigma_m \circ \sigma_l$ is not a reflection. So we're left with a rotation, a translation, or perhaps something we haven't yet encountered! The following proposition answers this question.

Proposition 160 *If lines l, m, and n are neither concurrent nor mutually parallel, the composition $\sigma_n \circ \sigma_m \circ \sigma_l$ is neither a reflection, a rotation, nor a translation.*

Proof. Let l, m, and n be three lines that are neither concurrent nor mutually parallel. Consider the composition $\alpha = \sigma_n \circ \sigma_m \circ \sigma_l$. By Proposition 159, α is not a reflection. It suffices to show that α is neither a rotation nor a translation.

<u>Case 1</u>: $l \cap m = C$. Then $\sigma_m \circ \sigma_l = \rho_{C,2\Theta}$ is a rotation of some angle $\Theta \notin 0°$ by Theorem 138. By substitution, $\alpha = \sigma_n \circ \rho_{C,2\Theta}$ so that $\sigma_n = \alpha \circ \rho_{C,-2\Theta}$. Suppose that α is either a rotation or a translation. Then by Theorem 157 (the Angle Addition Theorem), σ_n is either a rotation or a translation, which is a contradiction.

<u>Case 2</u>: $l \parallel m$. Then $\sigma_m \circ \sigma_l = \tau_{\mathbf{v}}$ is a translation by some vector $\mathbf{v}$ by Corollary 146. By substitution, $\alpha = \sigma_n \circ \tau_{\mathbf{v}}$ so that $\sigma_n = \alpha \circ \tau_{-\mathbf{v}}$. Suppose that α is either a rotation or a translation, Then by Theorem 157 (the Angle Addition Theorem), σ_n is either a rotation or a translation, which again is a contradiction.

In both cases, α is neither a rotation nor a translation. Therefore the composition $\sigma_n \circ \sigma_m \circ \sigma_l$ is neither a reflection, a rotation, nor a translation. ∎

Indeed, as Theorem 165 indicates, the composition $\sigma_n \circ \sigma_m \circ \sigma_l$ in Proposition 160 turns out to be a "glide reflection," which is an isometry unlike any we've encountered so far.

Definition 161 *Let c be a line and let $\mathbf{v}$ be a non-zero vector. A transformation $\gamma_{c,\mathbf{v}} : \mathbb{R}^2 \to \mathbb{R}^2$ is a **glide reflection** with **axis** c and **glide vector** $\mathbf{v}$ if*

1. $\gamma_{c,\mathbf{v}} = \sigma_c \circ \tau_{\mathbf{v}}$,

2. $\tau_{\mathbf{v}}(c) = c$.

*The **length** of $\gamma_{c,\mathbf{v}}$, denoted by $\|\gamma_{c,\mathbf{v}}\|$, is the norm $\|\mathbf{v}\|$.*

Glide reflections are isometries since translations and reflections are isometries by Theorems 103 and 130, and the composition of isometries is an isometry by Proposition 89.

Imagine the pattern of footprints you make when walking in the sand on a beach, and imagine that your footprint pattern extends infinitely far in either direction. Imagine a line c positioned midway between your left and right footprints. In your mind, slide the entire pattern one-half step in a direction parallel to c, then reflect it in line c. The image pattern, which exactly superimposes on the original pattern, is the result of performing a glide reflection with axis c (see Figure 5.15).

Figure 5.15. Footprints fixed by a glide reflection.

Here are some important properties of a glide reflection:

Proposition 162 *Let c be a line and let* **v** *be a non-zero vector.*

a. $\gamma_{c,\mathbf{v}}$ *interchanges the half-planes of c.*

b. $\gamma_{c,\mathbf{v}}$ *has no fixed points.*

c. *Let P be a point and let* $P' = \gamma_{c,\mathbf{v}}(P)$. *Then the midpoint of* $\overline{PP'}$ *is on c.*

d. $\gamma_{c,\mathbf{v}}$ *fixes exactly one line, its axis c.*

Proof. By definition, $\gamma_{c,\mathbf{v}} = \sigma_c \circ \tau_{\mathbf{v}}$. Let P be any point and let $Q = \tau_{\mathbf{v}}(P)$. If P is on c, so are Q and the midpoint M of $\overline{PQ}$, since $\tau_{\mathbf{v}}(c) = c$. Furthermore, $Q \neq P$ since $\mathbf{v} \neq \mathbf{0}$; thus $\gamma_{c,\mathbf{v}}(P) = \sigma_c(\tau_{\mathbf{v}}(P)) = \sigma_c(Q) = Q \neq P$ and $\gamma_{c,\mathbf{v}}$ has no fixed points on c. If P is off c, let $P' = \gamma_{c,\mathbf{v}}(P)$. Then Q and P lie on the same side of c, in which case P and P' lie on opposite sides of c. Thus $\gamma_{c,\mathbf{v}}$ interchanges the half-planes of c and has no fixed points off c. This proves (a) and (b).

Let $M = \overline{PP'} \cap c$, let R be the midpoint of $\overline{QP'}$, and let S be the foot of the perpendicular from M to $\overleftrightarrow{PQ}$ (see Figure 5.16). Then R lies on c by definition of σ_c, $\overline{MS} \cong \overline{RQ} \cong \overline{P'R}$, $\angle MPS \cong \angle P'MR$ since these angles are corresponding, and the angles $\angle P'RM$ and $\angle MSP$ are right angles. Therefore $\triangle P'RM \cong \triangle MSP$ by AAS and $PM = MP'$ (CPCTC) so that M is the midpoint of $\overline{PP'}$, which proves (c).

To prove (d), suppose $\gamma_{c,\mathbf{v}}(l) = l$ and let P be a point on l. Then $P' = \gamma_{c,\mathbf{v}}(P)$ is a point on $l = \overleftrightarrow{PP'}$ as is the midpoint M of $\overline{PP'}$. But M is also on c by part c above. Hence $M' = \gamma_{c,\mathbf{v}}(M)$ is on c by definition of $\gamma_{c,\mathbf{v}}$ and M' is on l by assumption. Therefore $l = c$ since M and M' are distinct points that both lie on l and c. ∎

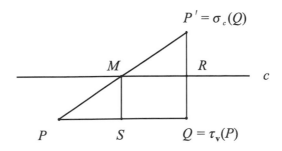

Figure 5.16. Proof of Proposition 162.

A glide reflection can be expressed as a composition of three reflections in the following way:

Theorem 163 *A transformation γ is a glide reflection with axis c if and only if there exist distinct parallels a and b perpendicular to c such that $\gamma = \sigma_c \circ \sigma_b \circ \sigma_a$.*

Proof. ($\Rightarrow$) Given a glide reflection γ with axis c, write $\gamma = \sigma_c \circ \tau$, where $\tau(c) = c$ and $\tau \neq \iota$. Let A be a point on c, then $A' = \tau(A)$ is also on c and is distinct from A. Let B be the midpoint of $\overline{AA'}$, and let a and b be the lines perpendicular to c at A and B, respectively. Then by Corollary 146, $\tau = \tau_{2AB} = \sigma_b \circ \sigma_a \neq \iota$ and it follows that $\gamma = \sigma_c \circ \sigma_b \circ \sigma_a$.

($\Leftarrow$) Given distinct parallel lines a and b, and a common perpendicular c, let $A = a \cap c$ and $B = b \cap c$. Then $\tau_{2AB}(c) = c$ and $\tau_{2AB} \neq \iota$ so that $\gamma = \sigma_c \circ \sigma_b \circ \sigma_a = \sigma_c \circ \tau_{2AB}$ is a glide reflection with axis c. ∎

A glide reflection can also be expressed as a reflection in some line l followed by a halfturn with center off l (or vice versa).

Theorem 164 *Let $\gamma : \mathbb{R}^2 \to \mathbb{R}^2$ be a transformation, let c be a line, and let $\mathbf{v}$ be a non-zero vector. The following are equivalent:*

a. *γ is a glide reflection with axis c and glide vector $\mathbf{v}$.*

b. *$\gamma = \sigma_c \circ \tau_{\mathbf{v}}$ and $\tau_{\mathbf{v}}(c) = c$.*

c. *There is a line $a \perp c$ and a point B on c and off a such that $\gamma = \varphi_B \circ \sigma_a$.*

d. *There is a line $b \perp c$ and a point A on c and off b such that $\gamma = \sigma_b \circ \varphi_A$.*

e. *$\gamma = \tau_{\mathbf{v}} \circ \sigma_c$ and $\tau_{\mathbf{v}}(c) = c$.*

Proof. Statements (a) and (b) are equivalent by definition.

(b)$\Rightarrow$(c) Given that $\gamma = \sigma_c \circ \tau_{\mathbf{v}}$ and $\tau_{\mathbf{v}}(c) = c$, use Theorem 163 to choose parallels a and b perpendicular to c such that $\tau_{\mathbf{v}} = \sigma_b \circ \sigma_a$, and let $B = b \cap c$. Then $\gamma = \sigma_c \circ \tau_{\mathbf{v}} = \sigma_c \circ \sigma_b \circ \sigma_a = \varphi_B \circ \sigma_a$ (see Figure 5.17).

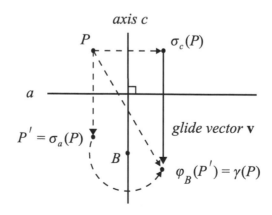

Figure 5.17. $\gamma = \varphi_B \circ \sigma_a$.

(c)$\Rightarrow$(d) Given a line $a \perp c$ and a point B on c and off a such that $\gamma = \varphi_B \circ \sigma_a$, let $A = a \cap c$ and let b be the line perpendicular to c through B. Then $\gamma = \varphi_B \circ \sigma_a = \sigma_c \circ \sigma_b \circ \sigma_a = \sigma_b \circ \sigma_c \circ \sigma_a = \sigma_b \circ \varphi_A$.

(d)$\Rightarrow$(e) Given a line $b \perp c$ and a point A on c and off b such that $\gamma = \sigma_b \circ \varphi_A$, let a be the line perpendicular to c through A. Then $\gamma = \sigma_b \circ \varphi_A = \sigma_b \circ \sigma_c \circ \sigma_a = \sigma_b \circ \sigma_a \circ \sigma_c = \tau_{\mathbf{v}} \circ \sigma_c$, and $\tau_{\mathbf{v}}(c) = c$ since c has direction $\mathbf{v}$.

(e)$\Rightarrow$(b) Given that $\gamma = \tau_{\mathbf{v}} \circ \sigma_c$ and $\tau_{\mathbf{v}}(c) = c$, use Theorem 163 to choose parallels a and b perpendicular to c such that $\tau_{\mathbf{v}} = \sigma_b \circ \sigma_a$. Then $\gamma = \tau_{\mathbf{v}} \circ \sigma_c = \sigma_b \circ \sigma_a \circ \sigma_c = \sigma_b \circ \sigma_c \circ \sigma_a = \sigma_c \circ \sigma_b \circ \sigma_a = \sigma_c \circ \tau_{\mathbf{v}}$. ∎

When a glide reflection γ is expressed as in Theorem 164, part c, its axis is the line through B perpendicular to a, and when γ is expressed as in Theorem 164, part d, its axis is the line through A and perpendicular to b.

Theorem 165 *Let l, m, and n be distinct lines. Then $\gamma = \sigma_n \circ \sigma_m \circ \sigma_l$ is a glide reflection if and only if l, m, and n are neither concurrent nor mutually parallel.*

Proof. ($\Rightarrow$) We prove the contrapositive. Suppose distinct lines l, m, and n are concurrent or mutually parallel. Then $\gamma = \sigma_n \circ \sigma_m \circ \sigma_l$ is a reflection by Corollary 158. But reflections have fixed points, and glide reflections have no fixed points by Proposition 162, part b. Therefore γ is not a glide reflection.

($\Leftarrow$) Assume l, m, and n are neither concurrent nor mutually parallel. Then l and m either intersect or are parallel.

Case 1: $l \cap m = P$. Then P is off n since l, m, and n are not concurrent. Let Q be the foot of the perpendicular from P to n, and let $q = \overleftrightarrow{PQ}$. By Theorem 141, there is a unique line p passing through P such that

$$\sigma_m \circ \sigma_l = \sigma_q \circ \sigma_p.$$

But $p \neq q$ since $l \neq m$, and Q is off p since $Q \neq P$ and Q is on q. Therefore, there is a point Q off line p such that

$$\gamma = \sigma_n \circ \sigma_m \circ \sigma_l = \sigma_n \circ \sigma_q \circ \sigma_p = \varphi_Q \circ \sigma_p,$$

which is a glide reflection by Theorem 164, part c.

Case 2: $l \parallel m$. Then n is a transversal for l and m since l, m, and n are not mutually parallel. Let $P = m \cap n$ and consider the composition $\sigma_l \circ \sigma_m \circ \sigma_n$. By the construction in Case 1 above, there is a unique line p passing through P and a point Q off p such that $\sigma_l \circ \sigma_m \circ \sigma_n = \varphi_Q \circ \sigma_p$. Then

$$\gamma = \sigma_n \circ \sigma_m \circ \sigma_l = (\sigma_l \circ \sigma_m \circ \sigma_n)^{-1} = (\varphi_Q \circ \sigma_p)^{-1} = \sigma_p \circ \varphi_Q$$

is a glide reflection by Theorem 164, part d. ∎

The equations of a glide reflection follow immediately from Theorem 164.

Corollary 166 *Let γ be a glide reflection with axis $l : aX + bY + c = 0$ with $a^2 + b^2 > 0$, and glide vector $\mathbf{v} = \begin{bmatrix} d \\ e \end{bmatrix}$. Then $ad + be = 0$ and the equations of γ are given by*

$$x' = x - \frac{2a}{a^2+b^2}(ax + by + c) + d$$
$$y' = y - \frac{2b}{a^2+b^2}(ax + by + c) + e.$$

Proof. Using Theorem 164 (e), write $\gamma = \tau_{\mathbf{v}} \circ \sigma_l$. Then $\tau_{\mathbf{v}}$ with vector $\mathbf{v} = \begin{bmatrix} d \\ e \end{bmatrix}$ fixes the axis $l : aX + bY + c = 0$ if and only if l is in the direction of $\mathbf{v}$ if and only if $ad + be = 0$. The equations of γ are the equations of the composition $\tau_{\mathbf{v}} \circ \sigma_l$. ∎

Example 167 Consider the line $m : 3X - 4Y + 1 = 0$ and the translation $\tau_{\mathbf{v}}$ where $\mathbf{v} = \begin{bmatrix} 4 \\ 3 \end{bmatrix}$. Since $ad + be = (3)(4) + (-4)(3) = 0$, the line m is in the direction of $\mathbf{v}$. Hence $\gamma = \sigma_m \circ \tau$ is the glide reflection with equations

$$x' = \tfrac{1}{25}(7x + 24y + 94)$$
$$y' = \tfrac{1}{25}(24x - 7y + 83).$$

Let $P = \begin{bmatrix} 0 \\ 19 \end{bmatrix}$ and $P' = \begin{bmatrix} 22 \\ -2 \end{bmatrix} = \gamma\left(\begin{bmatrix} 0 \\ 19 \end{bmatrix}\right)$; then the midpoint $M = \begin{bmatrix} 11 \\ \frac{17}{2} \end{bmatrix}$ of $\overline{PP'}$ is on m.

Here are some useful facts about glide reflections.

Theorem 168 *Let $\gamma_{c,\mathbf{v}}$ be a glide reflection.*

1. $\gamma_{c,\mathbf{v}}^{-1}$ *is a glide reflection with axis c and glide vector $-\mathbf{v}$; thus $\gamma_{c,\mathbf{v}}^{-1} = \gamma_{c,-\mathbf{v}}$.*

2. *If τ is any translation such that $\tau(c) = c$, then $\tau \circ \gamma_{c,\mathbf{v}} = \gamma_{c,\mathbf{v}} \circ \tau$.*

3. $\gamma_{c,\mathbf{v}}^2 = \tau_{2\mathbf{v}} \neq \iota$.

Proof. The proof of (a) is left to the reader. To prove (b), assume that τ is a non-trivial translation such that $\tau(c) = c$ and let A be a point on c. Then $B = \tau(A) \neq A$ is a point on c and $\tau = \tau_{\mathbf{AB}}$. Therefore $\sigma_c \circ \tau$ is a glide reflection with axis c and glide vector $\mathbf{AB}$, and

$$\sigma_c \circ \tau = \tau \circ \sigma_c \tag{5.7}$$

by Theorem 164, parts a, b, and e. On the other hand, $\gamma_{c,\mathbf{v}} = \sigma_c \circ \tau_{\mathbf{v}}$ by definition, and any two translations commute by Proposition 106, part 2. Therefore

$$\gamma_{c,\mathbf{v}} \circ \tau = \sigma_c \circ \tau_{\mathbf{v}} \circ \tau = \sigma_c \circ \tau \circ \tau_{\mathbf{v}} = \tau \circ \sigma_c \circ \tau_{\mathbf{v}} = \tau \circ \gamma_{c,\mathbf{v}}.$$

To prove (c), write $\gamma_{c,\mathbf{v}} = \sigma_c \circ \tau_{\mathbf{v}}$ and apply Theorem 164, parts a, b, and e to obtain

$$\gamma_{c,\mathbf{v}}^2 = (\sigma_c \circ \tau_{\mathbf{v}})^2 = \sigma_c \circ \tau_{\mathbf{v}} \circ \sigma_c \circ \tau_{\mathbf{v}} = \sigma_c \circ \sigma_c \circ \tau_{\mathbf{v}} \circ \tau_{\mathbf{v}} = \tau_{\mathbf{v}}^2 = \tau_{2\mathbf{v}}.$$

∎

We conclude this section with an important construction.

Theorem 169 *Let $\triangle ABC$ be a non-degenerate triangle and let a, b, and c be the lines containing the sides opposite A, B, and C. Then the axis and glide vector of the glide reflection $\gamma = \sigma_c \circ \sigma_b \circ \sigma_a$ is constructed in the following way:*

1. *If $\angle B$ is not a right angle, let P and Q be the feet of the altitudes on c and a, respectively, let t be the line through A perpendicular to $\overleftrightarrow{PQ}$, and let $R = \sigma_t (Q)$ (see Figure 5.18). Then $\overleftrightarrow{PQ}$ is the axis and $\mathbf{QR}$ is the glide vector.*

2. *If $\angle B$ is a right angle, let $t = \sigma_c (b)$ and let $R = \sigma_t (B)$. Then $\overleftrightarrow{BR}$ is the axis and $\mathbf{BR}$ is the glide vector.*

Proof. (1) If $\angle B$ is not a right angle, let $m = \overleftrightarrow{CP}$ and let l be the unique line passing through C such that $\sigma_b \circ \sigma_a = \sigma_m \circ \sigma_l$ given by Theorem 141, and write

$$\gamma = \sigma_c \circ \sigma_b \circ \sigma_a = \sigma_c \circ \sigma_m \circ \sigma_l = \varphi_P \circ \sigma_l.$$

Then by Theorem 164, part c, the axis of γ is the line through P perpendicular to l. Now repeat this argument at the vertex A: Let $s = \overleftrightarrow{AQ}$ and let t be the unique line passing through A such that $\sigma_c \circ \sigma_b = \sigma_t \circ \sigma_s$ and write

$$\gamma = \sigma_c \circ \sigma_b \circ \sigma_a = \sigma_t \circ \sigma_s \circ \sigma_a = \sigma_t \circ \varphi_Q.$$

Then by Theorem 164, part d, the axis of γ is the line through Q perpendicular to t. Finally, note that the points P and Q are distinct since $\angle B$ is not a right angle, and both are on the axis, since it is fixed by γ. Therefore $\overleftrightarrow{PQ}$ is the axis of γ. Furthermore, since the point $R = \gamma (Q) = (\sigma_t \circ \varphi_Q) (Q) = \sigma_t (Q)$ is also on the axis, $\mathbf{QR}$ is the glide vector of γ.

(2) If $\angle B$ is a right angle, let t be the unique line through A such that $\sigma_c \circ \sigma_b = \sigma_t \circ \sigma_c$ given by Theorem 141. Then composing both sides with σ_b gives $\sigma_c = \sigma_t \circ \sigma_c \circ \sigma_b$. To see that $\sigma_c (b) = t$, note that if P is a point on b, its image

$$P' = \sigma_c (P) = (\sigma_t \circ \sigma_c \circ \sigma_b) (P) = \sigma_t (\sigma_c (P)) = \sigma_t (P')$$

is a fixed point of σ_t, which is on t by definition of reflection. Hence $\sigma_c (b) = t$ since σ_c is a collineation. Now write

$$\gamma = \sigma_c \circ \sigma_b \circ \sigma_a = \sigma_t \circ \sigma_c \circ \sigma_a = \sigma_t \circ \varphi_B.$$

Then by Theorem 164, part d, the axis of γ is the line through B perpendicular to t. Furthermore, since $R = \gamma(B) = (\sigma_t \circ \varphi_B)(B) = \sigma_t(B)$ is also on the axis, $\overleftrightarrow{BR}$ is the axis, and $\mathbf{BR}$ is the glide vector of γ. ∎

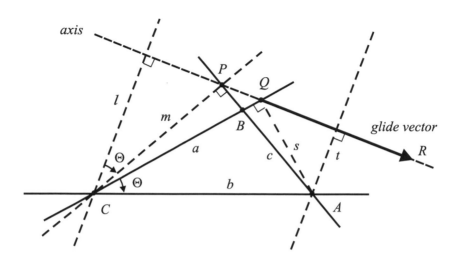

Figure 5.18. The axis and glide vector of $\gamma = \sigma_c \circ \sigma_b \circ \sigma_a$.

Exercises

1. The parallelogram "0" in the figure below is mapped to each of the other eight parallelograms by a reflection, a translation, a glide reflection or a halfturn. Indicate which of these apply in each case (more than one may apply in some cases).

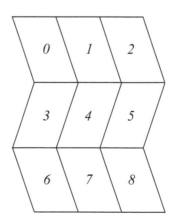

2. A glide reflection γ maps $\triangle ABC$ onto $\triangle A'B'C'$ in the diagram below. Use a MIRA to construct the axis and glide vector of γ.

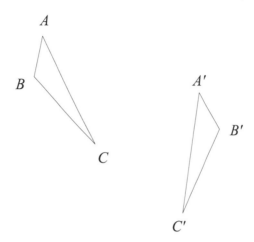

3. In the diagram below, parallels a and b are cut by transversal c. Use a MIRA and Theorem 164 to construct the axis and glide vector of the following glide reflections: $\gamma_1 = \sigma_c \circ \sigma_b \circ \sigma_a$; $\gamma_2 = \sigma_b \circ \sigma_c \circ \sigma_a$; $\gamma_3 = \sigma_c \circ \sigma_a \circ \sigma_b$; $\gamma_4 = \sigma_a \circ \sigma_c \circ \sigma_b$; $\gamma_5 = \sigma_b \circ \sigma_a \circ \sigma_c$; $\gamma_6 = \sigma_a \circ \sigma_b \circ \sigma_c$.

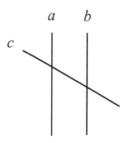

4. Consider a non-degenerate triangle $\triangle ABC$ with sides a, b, and c opposite vertices A, B, and C. Assume that $\triangle ABC$ is not a right triangle, and let P, Q, and R be the feet of the respective altitudes from C to c, from A to a, and from B to b. Use a MIRA and Theorem 169 to construct the axis and glide vector of the following glide reflections: $\gamma_1 = \sigma_c \circ \sigma_a \circ \sigma_b$; $\gamma_2 = \sigma_b \circ \sigma_c \circ \sigma_a$; $\gamma_3 = \sigma_a \circ \sigma_b \circ \sigma_c$.

5. Consider a non-degenerate right triangle $\triangle ABC$ with sides a, b, and c opposite vertices A, B, and C, and hypotenuse b. Use a MIRA and Theorem 169 to construct the axis and glide vector of the following glide reflections: $\gamma_1 = \sigma_c \circ \sigma_a \circ \sigma_b$; $\gamma_2 = \sigma_b \circ \sigma_c \circ \sigma_a$; $\gamma_3 = \sigma_a \circ \sigma_b \circ \sigma_c$.

6. Consider a non-degenerate $\triangle ABC$ with sides a, b, and c opposite vertices A, B, and C. Let $\gamma = \sigma_a \circ \sigma_b \circ \sigma_c$, and let Θ, Φ, and Ψ be the respective measures of the interior angles of $\triangle ABC$ indicated in the diagram below. Show that $\rho_{C,2\Psi} \circ \rho_{B,2\Phi} \circ \rho_{A,2\Theta} = \gamma^2 \neq \iota$ but $\rho_{A,2\Theta} \circ \rho_{B,2\Phi} \circ \rho_{C,2\Psi} = \iota$.

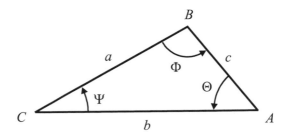

7. Prove the second half of Proposition 159: Given lines l, m, and n, and a line p such that $\sigma_p = \sigma_n \circ \sigma_m \circ \sigma_l$, prove that l, m, and n are either concurrent or mutually parallel.

8. If lines l, m, and n are either concurrent or mutually parallel, prove that $\sigma_n \circ \sigma_m \circ \sigma_l = \sigma_l \circ \sigma_m \circ \sigma_n$.

9. Given a non-degenerate triangle $\triangle ABC$, let b be the side opposite $\angle B$ and let l, m, and n be the respective angle bisectors of $\angle A$, $\angle B$, and $\angle C$. Then l, m, and n are concurrent by Exercise 3.2.17, and there is a unique line p such that $\sigma_n \circ \sigma_m \circ \sigma_l = \sigma_p$ by Corollary 158 (a). Prove that $p \perp b$.

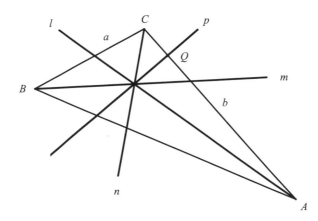

10. Let c be the line with equation $X - 2Y + 3 = 0$, let $P = \begin{bmatrix} 4 \\ -1 \end{bmatrix}$, and let $Q = \begin{bmatrix} 8 \\ 1 \end{bmatrix}$.

(a) Write the equations for the composite transformation $\gamma = \tau_{\mathbf{PQ}} \circ \sigma_c$.

(b) Find the image of $\begin{bmatrix} 1 \\ 2 \end{bmatrix}$, $\begin{bmatrix} -2 \\ 5 \end{bmatrix}$, and $\begin{bmatrix} -3 \\ -2 \end{bmatrix}$ under γ.

(c) Prove that $\gamma = \tau_{\mathbf{PQ}} \circ \sigma_c$ is a glide reflection.

11. Let γ be a glide reflection with axis c and glide vector $\mathbf{v}$. If $\gamma^2 = \tau_{2\mathbf{w}}$, show that $\mathbf{w} = \mathbf{v}$.

12. Let $\gamma_{c,\mathbf{v}}$ be a glide reflection with axis c and glide vector $\mathbf{v}$. Given any point M on c, construct a point P off c such that M is the midpoint of P and $\gamma_{c,\mathbf{v}}(P)$.

13. Given a translation $\tau_{\mathbf{v}}$, find the glide vector and axis of a glide reflection γ such that $\gamma^2 = \tau_{\mathbf{v}}$.

14. Prove Theorem 168 (a): If γ is a glide reflection with axis c and glide vector $\mathbf{v}$, then γ^{-1} is a glide reflection with axis c and glide vector $-\mathbf{v}$.

15. Use Theorem 165 to prove that the perpendicular bisectors of the sides in any non-degenerate triangle are concurrent (cf. Exercise 5.3.21).

Chapter 6

Classification of Isometries

Mathematics is the classification and study of all possible patterns.

Walter Warwick Sawyer (1911–2008)
Mathematician and Educator

So far we've encountered four families of isometries, namely, translations, rotations, reflections, and glide reflections, and you're probably wondering whether or not this list is complete. The answer to this question is the main result in this chapter.

6.1 The Fundamental Theorem and Congruence

In this section we prove the Fundamental Theorem of Transformational Geometry, which establishes the remarkable fact that an isometry can be expressed as a composition of three or fewer reflections. We apply the Fundamental Theorem to construct the (unique) isometry that sends one triangle to another congruent to it. But first we need some additional facts about fixed points.

Theorem 170 *An isometry that fixes two distinct points is either a reflection or the identity.*

Proof. We assume that $\alpha \neq \iota$ and show that α is a reflection. Let P and Q be distinct points, let $m = \overleftrightarrow{PQ}$, and let α be an isometry fixing P and Q. Let R be any point such that $R' = \alpha(R) \neq R$. Then P, Q, and R are non-collinear by Theorem 134 and $PR = PR'$ and $QR = QR'$ since α is an isometry. Since P and Q are equidistant from R and R', line m is the perpendicular bisector of $\overline{RR'}$. Hence P, Q, and R are non-collinear and

$$\alpha(R) = R' = \sigma_m(R), \quad \alpha(P) = P = \sigma_m(P), \quad \text{and} \quad \alpha(Q) = Q = \sigma_m(Q).$$

Therefore $\alpha = \sigma_m$ by Theorem 136 (the Three Points Theorem). ∎

Theorem 171 *An isometry that fixes exactly one point is a non-trivial rotation.*

Proof. Let α be an isometry with exactly one fixed point C, let P be a point distinct from C, and let $P' = \alpha(P)$. Since α is an isometry, $CP = CP'$ so that C is on the perpendicular bisector m of $\overline{PP'}$ (see Figure 6.1). But $\sigma_m(P') = P$ by definition of reflection, and it follows that

$$(\sigma_m \circ \alpha)(C) = \sigma_m(\alpha(C)) = \sigma_m(C) = C$$

and

$$(\sigma_m \circ \alpha)(P) = \sigma_m(\alpha(P)) = \sigma_m(P') = P.$$

Let $l = \overleftrightarrow{CP}$. Since $\sigma_m \circ \alpha$ is an isometry that fixes the distinct points C and P, either $\sigma_m \circ \alpha = \iota$ or $\sigma_m \circ \alpha = \sigma_l$ by Theorem 170. However, if $\sigma_m \circ \alpha = \iota$, then $\sigma_m = \alpha$, which is impossible since σ_m has infinitely many fixed points. Therefore $\sigma_m \circ \alpha = \sigma_l$ so that

$$\alpha = \sigma_m \circ \sigma_l.$$

Finally, $l \neq m$ since P is on l and off m. Hence α is a non-trivial rotation by Theorem 139. ∎

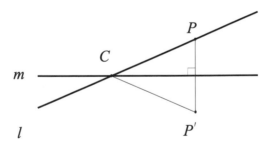

Figure 6.1. The fixed point C is on the perpendicular bisector of $\overline{PP'}$.

Thinking of the identity as a trivial rotation, we summarize Theorems 170 and 171 as:

Theorem 172 *An isometry with a fixed point is either a reflection, or a rotation. An isometry with exactly one fixed point is a non-trivial rotation.*

We have established the facts we need to prove our first major result, which characterizes an isometry as a composition of three or fewer reflections:

Theorem 173 (Fundamental Theorem of Transformational Plane Geometry) A transformation α is an isometry if and only if α can be expressed as a composition of three or fewer reflections.

Proof. ($\Rightarrow$) Let α be an isometry. If $\alpha = \iota$, choose any line l and write $\iota = \sigma_l \circ \sigma_l$. If $\alpha \neq \iota$, choose a point P such that $P' = \alpha(P) \neq P$ and let m be the perpendicular bisector of $\overline{PP'}$. Then

$$(\sigma_m \circ \alpha)(P) = \sigma_m(\alpha(P)) = \sigma_m(P') = P,$$

and $\beta = \sigma_m \circ \alpha$ fixes the point P. By Theorem 172, β is either a reflection or a rotation, the latter of which can be expressed as a composition of two reflections. Consequently, $\alpha = \sigma_m \circ \beta$ can be expressed as a composition of three or fewer reflections.

($\Leftarrow$) Consider a composition of reflections. Since reflections are isometries by Theorem 130, and the composition of isometries is an isometry by Proposition 89, a composition of reflections is an isometry. ∎

Our next theorem expresses triangle congruence in terms of isometries. Since an isometry is uniquely determined by its action on three non-collinear points by the Three Points Theorem (Theorem 136), the procedure used in the proof below also provides an alternative and constructive proof of the Fundamental Theorem (Theorem 173).

Theorem 174 *Triangles $\triangle PQR$ and $\triangle ABC$ are congruent if and only if there is an isometry α such that $\triangle ABC = \alpha(\triangle PQR)$. Furthermore, α is unique if and only if $\triangle PQR$ is scalene.*

Proof. ($\Rightarrow$) Given $\triangle PQR \cong \triangle ABC$, assume that $P \leftrightarrow A$, $Q \leftrightarrow B$, and $R \leftrightarrow C$ is a correspondence of vertices. By Theorem 136 (the Three Points Theorem), it is sufficient to define an isometry α with the desired property on the vertices of $\triangle PQR$. The following procedure produces such an isometry of the form $\alpha = \alpha_3 \circ \alpha_2 \circ \alpha_1$, where α_i is either the identity or a reflection: To begin, the SSS property of congruent triangles implies

$$AB = PQ, \quad AC = PR, \quad \text{and} \quad BC = QR \tag{6.1}$$

The isometry α_1: If $P = A$, let $\alpha_1 = \iota$. Otherwise, let $\alpha_1 = \sigma_l$ where l is the perpendicular bisector of $\overline{PA}$. In either case,

$$\alpha_1(P) = A.$$

Let

$$Q_1 = \alpha_1(Q) \quad \text{and} \quad R_1 = \alpha_1(R);$$

then

$$PQ = AQ_1, \quad PR = AR_1, \quad \text{and} \quad QR = Q_1R_1 \tag{6.2}$$

since the identity and a reflection are isometries (see Figure 6.2).

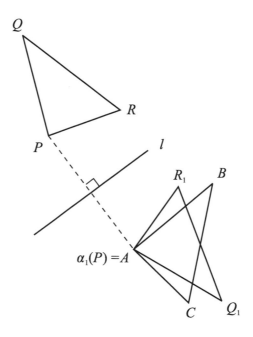

Figure 6.2. $\triangle AR_1Q_1 = \sigma_l(\triangle PQR)$.

The isometry α_2: If $Q_1 = B$, let $\alpha_2 = \iota$. Otherwise, let $\alpha_2 = \sigma_m$ where m is the perpendicular bisector of $\overline{Q_1B}$. By (6.1) and (6.2) we have

$$AB = PQ = AQ_1.$$

Then A is on m since A is equidistant from B and Q_1. In either case we have

$$\alpha_2(A) = A \quad \text{and} \quad \alpha_2(Q_1) = B.$$

Let

$$R_2 = \alpha_2(R_1)$$

and note that

$$AR_1 = AR_2 \quad \text{and} \quad Q_1R_1 = BR_2 \tag{6.3}$$

(see Figure 6.3).

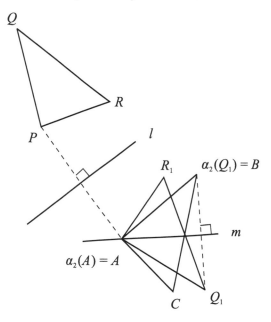

Figure 6.3. $\triangle ABC = \sigma_m(\triangle AQ_1R_1)$.

The isometry α_3: If $R_2 = C$, let $\alpha_3 = \iota$. Otherwise, let $\alpha_3 = \sigma_n$ where n is the perpendicular bisector of $\overline{R_2C}$. By (6.1), (6.2), and (6.3) we have

$$AC = PR = AR_1 = AR_2.$$

Then A is on n since A is equidistant from points C and R_2. On the other hand, (6.1), (6.2), and (6.3) also give

$$BC = QR = Q_1R_1 = BR_2.$$

Then B is on n since B is equidistant from points C and R_2. In either case we have

$$\alpha_3(A) = A, \quad \alpha_3(B) = B, \quad \text{and} \quad \alpha_3(R_2) = C.$$

Finally, set $\alpha = \alpha_3 \circ \alpha_2 \circ \alpha_1$ and note that

$$\alpha(P) = (\alpha_3 \circ \alpha_2 \circ \alpha_1)(P) = (\alpha_3 \circ \alpha_2)(A) = \alpha_3(A) = A;$$

$$\alpha(Q) = (\alpha_3 \circ \alpha_2 \circ \alpha_1)(Q) = (\alpha_3 \circ \alpha_2)(Q_1) = \alpha_3(B) = B;$$

$$\alpha(R) = (\alpha_3 \circ \alpha_2 \circ \alpha_1)(R) = (\alpha_3 \circ \alpha_2)(R_1) = \alpha_3(R_2) = C.$$

Furthermore, if $\triangle PQR$ is not scalene, it is isosceles, in which case there are multiple correspondences between the vertices of $\triangle PQR$ and $\triangle ABC$. Since the isometry α produced by the algorithm above depends on the choice of correspondence, and different correspondences produce distinct isometries, α is not unique.

($\Leftarrow$) Let α be an isometry such that $\alpha(P) = A$, $\alpha(Q) = B$ and $\alpha(R) = C$. Then $PQ = AB$, $PR = AC$, and $QR = BC$ by definition of isometry, and $\triangle PQR \cong \triangle ABC$ by SSS. If $\triangle PQR$ is scalene, the correspondence between the vertices of $\triangle PQR$ and $\triangle ABC$ is unique. Consequently, α is unique by the Three Points Theorem. ∎

The following remarkable characterization of congruent triangles is an immediate consequence of Theorems 173 and 174:

Corollary 175 $\triangle PQR \cong \triangle ABC$ *if and only if* $\triangle ABC$ *is the image of* $\triangle PQR$ *under a composition of three or fewer reflections.*

Corollary 176 *Two segments or two angles are congruent if and only if there exists an isometry mapping one onto the other.*

Proof. ($\Rightarrow$) Given two congruent segments or angles, construct a pair of congruent triangles containing them. Following the algorithm in the proof of Theorem 174, construct an isometry α from one triangle to the other. Then α sends one segment (or angle) to the other.

($\Leftarrow$) If α is an isometry that sends one segment (or angle) to another, the segments are congruent by definition (the angles are congruent by Proposition 92, part 3). ∎

Now we can define a general notion of congruence for arbitrary *plane figures*. Note that a plane figure is simply a non-empty subset of the plane.

Definition 177 *Two plane figures* s_1 *and* s_2 *are* **congruent** *if there is an isometry* α *such that* $s_2 = \alpha(s_1)$.

In this section we established the fact that two congruent triangles are related by a composition of three or fewer reflections. We identify this composition with a particular isometry in the next two sections.

Exercises

1. Let $P = \begin{bmatrix} 5 \\ 0 \end{bmatrix}$; $Q = \begin{bmatrix} 0 \\ 0 \end{bmatrix}$; $R = \begin{bmatrix} 0 \\ 10 \end{bmatrix}$. In each of the following, $\triangle PQR \cong \triangle ABC$. Apply the algorithm in the proof of Theorem 174 to find three or fewer lines such that the image of $\triangle PQR$ under reflections in these lines is $\triangle ABC$. If the composition is a

   - reflection, find the axis;
   - translation, find the translation vector;
   - rotation, find the center and angle of rotation;
   - glide reflection, find the axis and glide vector.

	A	B	C
(a)	$\begin{bmatrix} -5 \\ 0 \end{bmatrix}$	$\begin{bmatrix} 0 \\ 0 \end{bmatrix}$	$\begin{bmatrix} 0 \\ 10 \end{bmatrix}$
(b)	$\begin{bmatrix} 15 \\ 5 \end{bmatrix}$	$\begin{bmatrix} 10 \\ 5 \end{bmatrix}$	$\begin{bmatrix} 10 \\ 15 \end{bmatrix}$
(c)	$\begin{bmatrix} -5 \\ 0 \end{bmatrix}$	$\begin{bmatrix} 0 \\ 0 \end{bmatrix}$	$\begin{bmatrix} 0 \\ -10 \end{bmatrix}$
(d)	$\begin{bmatrix} 0 \\ 20 \end{bmatrix}$	$\begin{bmatrix} 0 \\ 15 \end{bmatrix}$	$\begin{bmatrix} -10 \\ 15 \end{bmatrix}$
(e)	$\begin{bmatrix} 10 \\ -10 \end{bmatrix}$	$\begin{bmatrix} 15 \\ -10 \end{bmatrix}$	$\begin{bmatrix} 15 \\ 0 \end{bmatrix}$

2. Let $P = \begin{bmatrix} 4 \\ 14 \end{bmatrix}$; $Q = \begin{bmatrix} 10 \\ 14 \end{bmatrix}$; $R = \begin{bmatrix} 10 \\ 6 \end{bmatrix}$. In each of the following, $\triangle PQR \cong \triangle ABC$. Apply the algorithm in the proof of Theorem 174 to find three or fewer lines such that the image of $\triangle PQR$ under reflections in these lines is $\triangle ABC$. If the composition is a

- reflection, find the axis;
- translation, find the translation vector;
- rotation, find the center and angle of rotation;
- glide reflection, find the axis and glide vector.

	A	B	C
(a)	$\begin{bmatrix} 14 \\ 24 \end{bmatrix}$	$\begin{bmatrix} 14 \\ 18 \end{bmatrix}$	$\begin{bmatrix} 22 \\ 18 \end{bmatrix}$
(b)	$\begin{bmatrix} 12 \\ 8 \end{bmatrix}$	$\begin{bmatrix} 18 \\ 8 \end{bmatrix}$	$\begin{bmatrix} 18 \\ 0 \end{bmatrix}$
(c)	$\begin{bmatrix} -4 \\ 14 \end{bmatrix}$	$\begin{bmatrix} -10 \\ 14 \end{bmatrix}$	$\begin{bmatrix} -10 \\ 22 \end{bmatrix}$
(d)	$\begin{bmatrix} -10 \\ 0 \end{bmatrix}$	$\begin{bmatrix} -10 \\ 6 \end{bmatrix}$	$\begin{bmatrix} -2 \\ 6 \end{bmatrix}$
(e)	$\begin{bmatrix} 6 \\ 0 \end{bmatrix}$	$\begin{bmatrix} 0 \\ 0 \end{bmatrix}$	$\begin{bmatrix} 0 \\ -8 \end{bmatrix}$

6.2   Classification of Isometries

In this section we give a complete solution of the classification problem for isometries (see Theorem 178). We also define the notions of "even" and "odd" isometries, i.e., compositions of an even or odd number of reflections, and observe that even isometries are translations or rotations and odd isometries are reflections or glide reflections (Theorem 182).

Theorem 178 (Classification of Isometries) *An isometry is exactly one of the following types: a reflection, a glide reflection, a rotation, or a non-trivial translation.*

Proof. We collect the various facts proved throughout our exposition.

- Every isometry is a composition of three or fewer reflections by Theorem 173.

- A composition of two reflections in distinct parallel lines is a non-trivial translation by Theorem 147.

- A composition of two reflections in distinct intersecting lines is a non-trivial rotation by Theorem 139.

- The identity is both a trivial translation and a trivial rotation.

- Non-trivial translations are fixed point free, but fix every line in the direction of translation by Theorem 110.

- By definition, a non-trivial rotation fixes exactly one point – its center – and a reflection fixes every point on its axis.

- Hence translations, non-trivial rotations, and reflections form mutually exclusive families of isometries.

- A composition of three reflections in concurrent or mutually parallel lines is a reflection by Corollary 158.

- A composition of three reflections in non-concurrent and non-mutually parallel lines is a glide reflection by Theorem 165.

- A glide reflection has no fixed points by Proposition 162, (b); consequently, a glide reflection is neither a rotation nor a reflection.

- A glide reflection fixes exactly one line – its axis – by Proposition 162, (d); consequently, a glide reflection is not a translation.

∎

The pictures in Figure 6.4 represent the various ways three or fewer lines can be configured. Successively reflecting in these lines gives us various ways to express the isometries in Theorem 178 as compositions of reflections. By reflecting exactly once in each line of the configuration, the first three represent reflections, the fourth and fifth represent glide reflections, the sixth represents a non-trivial rotation and the seventh represents a non-trivial translation. If we allow multiple reflections in the same line, compositions of the form $\sigma_a \circ \sigma_b \circ \sigma_a$ also represent reflections since the three axes of reflection are either parallel (when $a \parallel b$) or concurrent (when $a \nparallel b$).

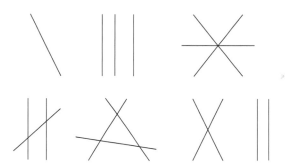

Figure 6.4. Configurations of lines representing reflections, translations, rotations, and glide reflections.

The Classification of Isometries is *mathematics par excellence* – a crown jewel of Transformational Plane Geometry. Indeed, the ultimate goal of every mathematical endeavor is to find and classify the objects studied. This is typically a profoundly difficult problem and rarely solved. Thus a complete solution of a classification problem calls for great celebration! And the Classification of Isometries is no exception.

Recall that the Fundamental Theorem (Theorem 173) asserts that an isometry can be expressed as a composition of three or fewer reflections. But how can one reduce a composition of four reflections (which is an isometry) to a composition of three or fewer reflections? Theorem 180 below answers this question. We begin with a lemma, which will be applied in the proof of Theorem 180.

Lemma 179 *Let a and b be lines and let P be a point. There exist lines c and d with c passing through P such that*

$$\sigma_d \circ \sigma_c = \sigma_b \circ \sigma_a.$$

Proof. If $a = b$, choose any line c passing through P and set $d = c$. Suppose $a \neq b$. If $a \cap b = P$, set $c = a$ and $d = b$. If $a \cap b = Q \neq P$, let $c = \overleftrightarrow{PQ}$. Then by Theorem 141, there is a unique line d such that $\sigma_d \circ \sigma_c = \sigma_b \circ \sigma_a$. If $a \parallel b$, let c be the line through P parallel to a. Then by Theorem 149, there is a unique line d such that $\sigma_d \circ \sigma_c = \sigma_b \circ \sigma_a$. ∎

Theorem 180 *A composition of four reflections reduces to a composition of two reflections, i.e., given lines p, q, r, and s, there exist lines l and m such that*

$$\sigma_s \circ \sigma_r \circ \sigma_q \circ \sigma_p = \sigma_m \circ \sigma_l.$$

Proof. Choose a point P on line p and consider the composition $\sigma_r \circ \sigma_q$ (see Figure 6.5). By Lemma 179, there exist lines q_1 and r_1 with q_1 passing through P such that

$$\sigma_{r_1} \circ \sigma_{q_1} = \sigma_r \circ \sigma_q.$$

Next, consider the composition $\sigma_s \circ \sigma_{r_1}$. By Lemma 179, there exist lines r_2 and m with r_2 passing through P such that

$$\sigma_m \circ \sigma_{r_2} = \sigma_s \circ \sigma_{r_1}.$$

Since p, q_1, and r_2 are concurrent at P, there is a unique line l such that

$$\sigma_{r_2} \circ \sigma_{q_1} \circ \sigma_p = \sigma_l$$

by Corollary 158. Therefore

$$\sigma_s \circ \sigma_r \circ \sigma_q \circ \sigma_p = \sigma_s \circ \sigma_{r_1} \circ \sigma_{q_1} \circ \sigma_p = \sigma_m \circ \sigma_{r_2} \circ \sigma_{q_1} \circ \sigma_p = \sigma_m \circ \sigma_l.$$

∎

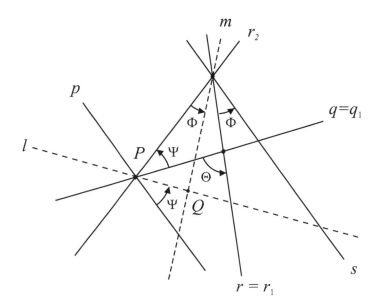

Figure 6.5. Here we chose $P = p \cap q$.

Repeated applications of Theorem 180 reduce a composition of an odd number of reflections to three reflections or one, but never to a composition of two. Likewise, a composition of an even number of reflections reduces to two reflections or the identity, but never to a composition of three or one. Since every isometry is a composition of three or fewer reflections (the Fundamental Theorem), each isometry falls into exactly one of two mutually exclusive families: (1) isometries that can be expressed as a composition of an even number of reflections, and (2) isometries that can be expressed as a composition of an odd number of reflections.

Definition 181 *An isometry α is **even** if α can be expressed as a composition of an even number of reflections; otherwise α is **odd**.*

Theorem 182 *An isometry α is even if and only if α is a translation or a rotation (the identity is a trivial rotation); α is odd if and only if α is a reflection or a glide reflection.*

Proof. Let α be an isometry. By Theorem 178 (Classification of Isometries), α is exactly one of the following types: a reflection, a glide reflection, a rotation, or a non-trivial translation. It suffices to gather together various facts proved throughout our exposition.

- If α is a rotation (the identity is a trivial rotation), then α is even by Theorem 139.

- If α is a non-trivial translation, then α is even by Theorem 147.

- If α is a reflection, then α is odd by definition.

- If α is a glide reflection, then α is odd by Theorem 165.

∎

Theorem 183 *An even involutory isometry is a halfturn; an odd involutory isometry is a reflection.*

Proof. Reflections are involutory isometries, so consider an involutory isometry α that is not a reflection. We claim that α is a halfturn. Since an involution is not the identity, there exist distinct points P and Q such that $Q = \alpha(P)$. By composing α with both sides we obtain

$$\alpha(Q) = \alpha^2(P) = \iota(P) = P.$$

Hence α interchanges the points P and Q. Let M be the midpoint of P and Q, then $P - M - Q$ and $PM = MQ$. Let $M' = \alpha(M)$; then $P - M' - Q$ and $PM' = M'Q$ since α is an isometry, a collineation, and preserves betweenness. Therefore $M' = M$ and M is a fixed point. Therefore α is either a non-trivial rotation about M or a reflection in some line containing M by Theorem

172. Since α is not a reflection, it is an involutory rotation about M, i.e., a halfturn by Corollary 126. Therefore $\alpha = \varphi_M$, and it follows that an involutory isometry is either a reflection or a halfturn. The conclusion now follows from the fact that reflections are odd and halfturns are even. ∎

Exercises

In Exercises 1–6, the respective equations of lines p, q, r, and s are given. In each case, find lines l and m such that $\sigma_s \circ \sigma_r \circ \sigma_q \circ \sigma_p = \sigma_m \circ \sigma_l$ and identify the isometry $\sigma_m \circ \sigma_l$ as a translation, a rotation, or the identity.

1. $Y = X$, $Y = X + 1$, $Y = X + 2$, and $Y = X + 3$.

2. $Y = X$, $Y = X + 1$, $Y = X + 2$, and $Y = -X$.

3. $X = 0$, $Y = 0$, $Y = X$, and $Y = -X$.

4. $X = 0$, $Y = 0$, $Y = 2$, and $X = 2$.

5. $X = 0$, $Y = 0$, $X = 1$, and $Y = X + 2$.

6. $X = 0$, $Y = 0$, $Y = X + 1$, and $Y = -X + 2$.

7. Prove that an even isometry with two distinct fixed points is the identity.

8. Identify the isometric dilatations and prove your answer.

9. In the diagram below, use a MIRA to construct lines l and m such that $\sigma_s \circ \sigma_r \circ \sigma_q \circ \sigma_p = \sigma_m \circ \sigma_l$.

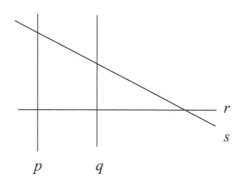

10. Consider the following diagram:

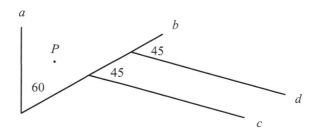

(a) Use a MIRA to construct the point Q such that $\rho_{Q,\Theta} = \sigma_b \circ \varphi_P \circ \sigma_a$ and find the rotation angle Θ.

(b) Use a MIRA to construct the point R such that $\rho_{R,\Phi} = \sigma_d \circ \sigma_c \circ \sigma_b \circ \sigma_a$ and find the rotation angle Φ.

6.3   Orientation and the Isometry Recognition Problem

Given two congruent triangles, the procedure in the proof of Theorem 174 gives us a constructive way to identify the isometry that relates them as a composition of three or fewer reflections. By Theorem 178 (Classification of Isometries), the isometry that relates the two congruent triangles is exactly one of the following types: a reflection, a glide reflection, a rotation, or a non-trivial translation. But how can one determine at a glance which of the four types of isometries it is? Our goal for this section is to solve this *Isometry Recognition Problem*.

A choice of "orientation" of the plane determines which way we measure positive angles – clockwise or counterclockwise. Given the "standard" orientation, positive angles are measured counterclockwise, but the opposite alternative would serve equally well. As we shall see, even isometries preserve orientation while odd isometries reverse it. The precise definition of an orientation requires some facts from linear algebra.

Let $\mathbf{a} = \begin{bmatrix} a_1 \\ a_2 \end{bmatrix}$ and $\mathbf{b} = \begin{bmatrix} b_1 \\ b_2 \end{bmatrix}$ be vectors. The *determinant of the matrix* $[\mathbf{a} \mid \mathbf{b}] = \begin{bmatrix} a_1 & b_1 \\ a_2 & b_2 \end{bmatrix}$ is defined to be

$$\det \begin{bmatrix} a_1 & b_1 \\ a_2 & b_2 \end{bmatrix} := a_1 b_2 - a_2 b_1.$$

Note that

$$\det [\mathbf{a} \mid \mathbf{b}] = -\det [\mathbf{b} \mid \mathbf{a}].$$

A *basis for* $\mathbb{R}^2$ is a pair of non-zero, non-parallel vectors $\{\mathbf{v}_1, \mathbf{v}_2\}$. Thus $\{\mathbf{v}_1, \mathbf{v}_2\}$ is a basis if and only if

$$\det [\mathbf{v}_1 \mid \mathbf{v}_2] < 0 \quad \text{or} \quad \det [\mathbf{v}_1 \mid \mathbf{v}_2] > 0.$$

The sign of the determinant is an equivalence relation on the set of all *ordered bases* for $\mathbb{R}^2$ and places a given ordered basis $\{\mathbf{v}_1, \mathbf{v}_2\}$ in one of two equivalence classes – bases with negative determinant or bases with positive determinant.

Definition 184 *An **orientation** of $\mathbb{R}^2$ is a choice of an ordered basis $\{\mathbf{v}_1, \mathbf{v}_2\}$. An orientation $\{\mathbf{v}_1, \mathbf{v}_2\}$ is **positive** (respectively, **negative**) if $\det [\mathbf{v}_1 \mid \mathbf{v}_2] > 0$ (respectively, $\det [\mathbf{v}_1 \mid \mathbf{v}_2] < 0$).*

The following theorem asserts that the orientation $\{\mathbf{v}_1, \mathbf{v}_2\}$ is positive if and only if the measure of the angle from $\mathbf{v}_1$ to $\mathbf{v}_2$ is positive.

Theorem 185 *Let $\{\mathbf{v}_1, \mathbf{v}_2\}$ be an orientation of $\mathbb{R}^2$ and let Θ be the measure of the angle from $\mathbf{v}_1$ to $\mathbf{v}_2$. Then $\{\mathbf{v}_1, \mathbf{v}_2\}$ is positive if and only if $\Theta > 0$.*

Proof. Let Θ_i be the measure of the angle from $\mathbf{e}_1 = \begin{bmatrix} 1 \\ 0 \end{bmatrix}$ to $\mathbf{v}_i$ $(i = 1, 2)$. Then $\Theta \equiv \Theta_2 - \Theta_1$, $\mathbf{v}_1 = \begin{bmatrix} \|\mathbf{v}_1\| \cos \Theta_1 \\ \|\mathbf{v}_1\| \sin \Theta_1 \end{bmatrix}$, $\mathbf{v}_2 = \begin{bmatrix} \|\mathbf{v}_2\| \cos \Theta_2 \\ \|\mathbf{v}_2\| \sin \Theta_2 \end{bmatrix}$, and

$$\begin{aligned} \det [\mathbf{v}_1 | \mathbf{v}_2] &= (\|\mathbf{v}_1\| \cos \Theta_1) (\|\mathbf{v}_2\| \sin \Theta_2) - (\|\mathbf{v}_1\| \sin \Theta_1) (\|\mathbf{v}_2\| \cos \Theta_2) \\ &= \|\mathbf{v}_1\| \, \|\mathbf{v}_2\| \, (\cos \Theta_1 \sin \Theta_2 - \sin \Theta_1 \cos \Theta_2) \\ &= \|\mathbf{v}_1\| \, \|\mathbf{v}_2\| \sin (\Theta_2 - \Theta_1) = \|\mathbf{v}_1\| \, \|\mathbf{v}_2\| \sin \Theta. \end{aligned}$$

Note that the measure Θ of the angle from $\mathbf{v}_1$ to $\mathbf{v}_2$ lies in $(-180, 180]$ (see Figure 6.6). By definition, the orientation $\{\mathbf{v}_1, \mathbf{v}_2\}$ is positive if and only if $\det [\mathbf{v}_1 | \mathbf{v}_2] = \|\mathbf{v}_1\| \, \|\mathbf{v}_2\| \sin \Theta > 0$ if and only if $\sin \Theta > 0$ if and only if $\Theta > 0$. ∎

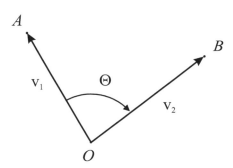

Figure 6.6. The orientation $\{\mathbf{v}_1, \mathbf{v}_2\}$ is negative.

Example 186 The standard ordered basis $\{\mathbf{e}_1, \mathbf{e}_2\}$ is a positive orientation of $\mathbb{R}^2$ and the measure of the angle from $\mathbf{e}_1$ to $\mathbf{e}_2$ is positive.

Definition 187 *Let α be a transformation with the following property: If $\{\mathbf{v}_1, \mathbf{v}_2\}$ is an orientation of $\mathbb{R}^2$, so is $\{\alpha(\mathbf{v}_1), \alpha(\mathbf{v}_2)\}$. Then α is* **orientation-preserving** *if the orientations $\{\mathbf{v}_1, \mathbf{v}_2\}$ and $\{\alpha(\mathbf{v}_1), \alpha(\mathbf{v}_2)\}$ are both positive or both negative; otherwise α is* **orientation-reversing.**

Since a given isometry α is a composition of reflections, determining whether α preserves or reverses orientation is immediate once this is known for reflections. Thus our next proposition is crucial.

Proposition 188 *Reflections are orientation-reversing.*

Proof. Let $\{\mathbf{v}_1, \mathbf{v}_2\}$ be an orientation; then $\det[\mathbf{v}_1 \mid \mathbf{v}_2] \neq 0$. A general line m has equation $aX + bY + c = 0$ with $a^2 + b^2 > 0$. The reflection of a vector $\mathbf{v} = \begin{bmatrix} x \\ y \end{bmatrix}$ in line m is given by Equation (4.8), i.e.,

$$\sigma_m(\mathbf{v}) = \begin{bmatrix} x - \frac{2a}{a^2+b^2}(ax + by) \\ y - \frac{2b}{a^2+b^2}(ax + by) \end{bmatrix}.$$

We may express this equation in matrix form

$$\sigma_m(\mathbf{v}) = A\mathbf{v},$$

where

$$A = \frac{1}{a^2 + b^2}\begin{bmatrix} b^2 - a^2 & -2ab \\ -2ab & a^2 - b^2 \end{bmatrix}.$$

By a straightforward computation,

$$\det A = \frac{1}{(a^2 + b^2)^2}\det\begin{bmatrix} b^2 - a^2 & -2ab \\ -2ab & a^2 - b^2 \end{bmatrix} = \frac{-\left(a^2 - b^2\right)^2 - 4a^2b^2}{(a^2 + b^2)^2} = -1.$$

Therefore

$$\det[\sigma_m(\mathbf{v}_1) \mid \sigma_m(\mathbf{v}_2)] = \det(A[\mathbf{v}_1 \mid \mathbf{v}_2]) = (\det A)(\det[\mathbf{v}_1 \mid \mathbf{v}_2]) = -\det[\mathbf{v}_1 \mid \mathbf{v}_2].$$

Since $\det[\sigma_m(\mathbf{v}_1) \mid \sigma_l(\mathbf{v}_2)]$ and $\det[\mathbf{v}_1 \mid \mathbf{v}_2]$ have opposite signs, σ_m is orientation-reversing by definition. ∎

Theorem 189 *An isometry α is orientation-preserving if and only if α is even (a translation or a rotation); α is orientation-reversing if and only if α is odd (a reflection or a glide reflection).*

Proof. Let α be an isometry. Since a reflection reverses orientation by Proposition 188, a composition of an even number of reflections preserves orientation while a composition of an odd number of reflections reverses it. Thus α is even if and only if α is a composition of an even number of reflections if and only if α preserves orientation. Since translations and rotations are even, and reflections and glide reflections are odd by Theorem 182, the proof is complete. ∎

We now consider the Isometry Recognition Problem posed above. Let α be an isometry, let A, B, and C be non-collinear points, let $A' = \alpha(A)$, $B' = \alpha(B)$, and $C' = \alpha(C)$, and recall that α preserves angle measurement up to sign, i.e., $m\angle A'B'C' = \pm m\angle ABC$ (Proposition 92, part 3). If α is a translation or a rotation, it seems intuitively obvious that $m\angle A'B'C' = m\angle ABC$. Likewise, if α is a reflection or a glide reflection, it seems intuitively obvious that $m\angle A'B'C' = -m\angle ABC$. Indeed, this is the essence of Theorem 191, which follows our next definition.

Definition 190 *An isometry α is* ***direct*** *if $m\angle A'B'C' = m\angle ABC$ for all non-collinear points A, B, and C, and their images $A' = \alpha(A)$, $B' = \alpha(B)$, and $C' = \alpha(C)$; α is* ***opposite*** *if $m\angle A'B'C' = -m\angle ABC$.*

When the isometry α in Definition 190 is direct, both $\angle ABC$ and $\angle A'B'C'$ are measured clockwise or both are measured counterclockwise. When α is opposite, either $\angle ABC$ is measured counterclockwise and $\angle A'B'C'$ is measured clockwise, or vice versa. Thus if $\triangle ABC \cong \triangle A'B'C'$, the isometry α sending $\triangle ABC$ to $\triangle A'B'C'$ is direct if $m\angle A'B'C' = m\angle ABC$ and is opposite if $m\angle A'B'C' = -m\angle ABC$. This observation together with our next Theorem makes it easy to determine the isometry type of α at a glance.

Theorem 191 *An isometry α is direct if and only if α is even (a translation or a rotation); α is opposite if and only if α is odd (a reflection or a glide reflection).*

Proof. Let α be an isometry, let A, B, and C be non-collinear points, and let $A' = \alpha(A)$, $B' = \alpha(B)$, and $C' = \alpha(C)$. Consider $\mathbf{v}_1 = \mathbf{BA}$ and $\mathbf{v}_2 = \mathbf{BC}$. Then $m\angle ABC$ is the measure of the angle from $\mathbf{v}_1$ to $\mathbf{v}_2$. By definition, α is direct if and only if $m\angle A'B'C' = m\angle ABC$. By Theorem 185, $m\angle ABC > 0$ if and only if the orientation $\{\mathbf{v}_1, \mathbf{v}_2\}$ is positive. Thus $m\angle A'B'C' = m\angle ABC$ if and only if the orientations $\{\mathbf{v}_1, \mathbf{v}_2\}$ and $\{\alpha(\mathbf{v}_1), \alpha(\mathbf{v}_2)\}$ are both positive or both negative. An application of Theorem 189 completes the proof. ∎

The results in Theorems 182, 189, and 191 are summarized in Theorem 192; the rotation and glide reflection pictured in Figure 6.7 confirm the conclusion.

Theorem 192 *Translations and rotations are even, direct, and orientation-preserving; reflections and glide reflections are odd, opposite, and orientation-reversing.*

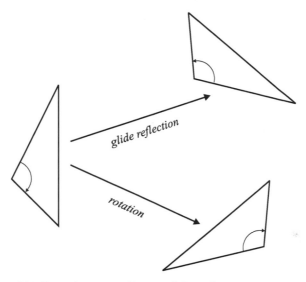

Figure 6.7. Rotations are direct; glide reflections are opposite.

Exercises

1. Determine whether the ordered basis $\{v_1, v_2\}$ is a positive or a negative orientation.

   (a) $v_1 = \begin{bmatrix} 8 \\ 9 \end{bmatrix}$ and $v_2 = \begin{bmatrix} 7 \\ 8 \end{bmatrix}$.

   (b) $v_1 = \begin{bmatrix} 6 \\ 6 \end{bmatrix}$ and $v_2 = \begin{bmatrix} -5 \\ -7 \end{bmatrix}$.

2. Let $A = \begin{bmatrix} 4 \\ 14 \end{bmatrix}$; $B = \begin{bmatrix} 10 \\ 14 \end{bmatrix}$; $C = \begin{bmatrix} 10 \\ 6 \end{bmatrix}$. In each of the following, $\triangle ABC \cong \triangle A'B'C'$ and α is the unique isometry such that $\alpha(A) = A'$, $\alpha(B) = B'$, and $\alpha(C) = C'$. Determine by inspection whether α preserves or reverses orientation. Do not apply the algorithm in the proof of Theorem 174. Compare with Exercise 6.1.2.

	A'	B'	C'
(a)	$\begin{bmatrix} 14 \\ 24 \end{bmatrix}$	$\begin{bmatrix} 14 \\ 18 \end{bmatrix}$	$\begin{bmatrix} 22 \\ 18 \end{bmatrix}$
(b)	$\begin{bmatrix} 12 \\ 8 \end{bmatrix}$	$\begin{bmatrix} 18 \\ 8 \end{bmatrix}$	$\begin{bmatrix} 18 \\ 0 \end{bmatrix}$
(c)	$\begin{bmatrix} -4 \\ 14 \end{bmatrix}$	$\begin{bmatrix} -10 \\ 14 \end{bmatrix}$	$\begin{bmatrix} -10 \\ 22 \end{bmatrix}$
(d)	$\begin{bmatrix} -10 \\ 0 \end{bmatrix}$	$\begin{bmatrix} -10 \\ 6 \end{bmatrix}$	$\begin{bmatrix} -2 \\ 6 \end{bmatrix}$
(e)	$\begin{bmatrix} 6 \\ 0 \end{bmatrix}$	$\begin{bmatrix} 0 \\ 0 \end{bmatrix}$	$\begin{bmatrix} 0 \\ -8 \end{bmatrix}$

3. The six triangles in the diagram below are congruent. Name the isometry that maps $\triangle ABC$ to the five triangles (a) through (e) and state whether the isometry is direct or opposite. If the isometry is a rotation, determine the rotation angle.

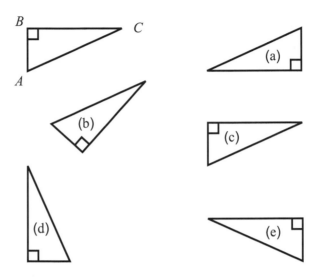

4. Prove directly as in the proof of Proposition 188 that translations are orientation-preserving.

5. Prove directly as in the proof of Proposition 188 that rotations are orientation-preserving.

6. Prove directly as in the proof of Proposition 188 that glide reflections are orientation-reversing.

6.4   The Geometry of Conjugation

In this section we give a geometrical interpretation of algebraic "conjugation." Conjugation is fundamentally important in geometry because it preserves all properties of isometries up to sign of rotation angle. For example, the conjugate of a translation is a translation and the norms of their glide vectors are equal. We use conjugation to relate a given isometry to a conjugate with particularly nice properties. For example, a general rotation is related to a conjugate rotation about the origin with comparatively simple equations (see Example 194).

Recall that when rationalizing the denominator of $\frac{1}{2+\sqrt{3}}$ we *conjugate by* $2 - \sqrt{3}$ as follows:

$$\frac{1}{2 + \sqrt{3}} = \left(\frac{1}{2 + \sqrt{3}}\right)\left(\frac{2 - \sqrt{3}}{2 - \sqrt{3}}\right) = 2 - \sqrt{3}.$$

Since *multiplication of real numbers is commutative*, we can express this conjugation in the following seemingly awkward way:

$$\left(\frac{1}{2 + \sqrt{3}}\right)\left(\frac{2 - \sqrt{3}}{2 - \sqrt{3}}\right) = \left(2 - \sqrt{3}\right)\left(\frac{1}{2 + \sqrt{3}}\right)\left(2 - \sqrt{3}\right)^{-1}. \qquad (6.4)$$

But when multiplication is non-commutative, as it is when composing isometries for example, conjugation always takes the form in the right-hand-side of Equation (6.4).

Definition 193 *Let α and β be isometries. The **conjugate of β by α** is the composition $\alpha \circ \beta \circ \alpha^{-1}$.*

Example 194 A general rotation $\rho_{C,\Theta}$ can be expressed as the conjugate of $\rho_{O,\Theta}$ by τ_{OC}, i.e.,

$$\rho_{C,\Theta} = \tau_{OC} \circ \rho_{O,\Theta} \circ \tau_{OC}^{-1}. \qquad (6.5)$$

Since the equations of $\rho_{O,\Theta}$, τ_{OC}, and τ_{OC}^{-1} are relatively simple, it's easy to compose them and obtain the equations of $\rho_{C,\Theta}$ as we did in the proof of Theorem 114.

Example 195 Note that $\beta = \alpha \circ \beta \circ \alpha^{-1}$ if and only if $\alpha \circ \beta = \beta \circ \alpha$ if and only if α and β commute. Thus, given any two vectors **u** and **v**, the calculation

$$\tau_{\mathbf{v}} \circ \tau_{\mathbf{u}} \circ \tau_{\mathbf{v}}^{-1} = \tau_{\mathbf{v}} \circ \tau_{\mathbf{u}} \circ \tau_{-\mathbf{v}} = \tau_{\mathbf{v}+\mathbf{u}-\mathbf{v}} = \tau_{\mathbf{u}}$$

provides a simple proof that translations commute.

Here are some algebraic properties of conjugation.

Theorem 196 *a. The square of a conjugate is the conjugate of the square.*

b. The conjugate of an involution is an involution.

Proof. Let α and β be isometries.

(a) $\left(\alpha \circ \beta \circ \alpha^{-1}\right)^2 = \alpha \circ \beta \circ \alpha^{-1} \circ \alpha \circ \beta \circ \alpha^{-1} = \alpha \circ \beta^2 \circ \alpha^{-1}$.

(b) Let β be an involution. First note that $\alpha \circ \beta \circ \alpha^{-1} = \iota$ if and only if $\beta = \iota$. Thus $\beta \neq \iota$ implies $\alpha \circ \beta \circ \alpha^{-1} \neq \iota$. On the other hand, by part a we have $\left(\alpha \circ \beta \circ \alpha^{-1}\right)^2 = \alpha \circ \beta^2 \circ \alpha^{-1} = \alpha \circ \iota \circ \alpha^{-1} = \alpha \circ \alpha^{-1} = \iota$. Therefore $\alpha \circ \beta \circ \alpha^{-1}$ is an involution. $\blacksquare$

Here are some geometrical consequences.

Proposition 197 *Conjugation preserves parity, i.e., if α and β are isometries, then β and $\alpha \circ \beta \circ \alpha^{-1}$ have the same parity. Thus β and $\alpha \circ \beta \circ \alpha^{-1}$ are both direct or are both opposite.*

Proof. By Theorem 173 we know that α can be expressed as a composition of reflections. Since the inverse of a composition is the composition of the inverses in reverse order, α and α^{-1} have the same parity and together contribute an even number of factors to every factorization of $\alpha \circ \beta \circ \alpha^{-1}$ as a composition of reflections. Hence β and $\alpha \circ \beta \circ \alpha^{-1}$ have the same parity. The fact that β and $\alpha \circ \beta \circ \alpha^{-1}$ both are both direct or are both opposite follows from Theorem 191. ∎

Theorem 198 *Let α be an isometry.*

    a. *The conjugate of a halfturn is a halfturn. In fact, if P is any point, then*

$$\alpha \circ \varphi_P \circ \alpha^{-1} = \varphi_{\alpha(P)}.$$

    b. *The conjugate of a reflection is a reflection. In fact, if m is any line, then*

$$\alpha \circ \sigma_m \circ \alpha^{-1} = \sigma_{\alpha(m)}.$$

Proof. (a) Since φ_P is even by Theorem 182, so is $\alpha \circ \varphi_P \circ \alpha^{-1}$ by Proposition 197. Furthermore, $\alpha \circ \varphi_P \circ \alpha^{-1}$ is an involution by Theorem 196. By Theorem 183, involutory isometries are either reflections (which are odd) or halfturns (which are even), so $\alpha \circ \varphi_P \circ \alpha^{-1}$ is a halfturn. To locate its center, observe that

$$\left(\alpha \circ \varphi_P \circ \alpha^{-1}\right)(\alpha(P)) = \left(\alpha \circ \varphi_P \circ \alpha^{-1} \circ \alpha\right)(P) = (\alpha \circ \varphi_P)(P) = \alpha(P).$$

Thus $\alpha(P)$ is fixed by the halfturn $\alpha \circ \varphi_P \circ \alpha^{-1}$. Therefore $\alpha(P)$ is the center of the halfturn $\alpha \circ \varphi_P \circ \alpha^{-1}$ and we have $\alpha \circ \varphi_P \circ \alpha^{-1} = \varphi_{\alpha(P)}$. The conjugation of halfturn by a translation is pictured in Figure 6.8.

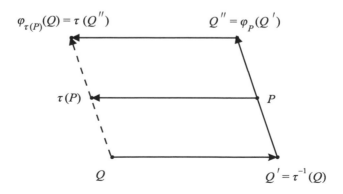

Figure 6.8. Conjugation of φ_P by τ.

(b) Since σ_m is odd, so is $\alpha \circ \sigma_m \circ \alpha^{-1}$ by Proposition 197. Furthermore $\alpha \circ \sigma_m \circ \alpha^{-1}$ is an involution by Theorem 196. By Theorem 183 involutory isometries are either halfturns (which are even) or reflections (which are odd), so $\alpha \circ \sigma_m \circ \alpha^{-1}$ is a reflection. To determine its axis, observe that if P is any point on m, then

$$\left(\alpha \circ \sigma_m \circ \alpha^{-1}\right)(\alpha(P)) = \left(\alpha \circ \sigma_m \circ \alpha^{-1} \circ \alpha\right)(P) = (\alpha \circ \sigma_m)(P) = \alpha(P).$$

Thus every point $\alpha(P)$ on $\alpha(m)$ is fixed by the reflection $\alpha \circ \sigma_m \circ \alpha^{-1}$. Therefore $\alpha(m)$ is the axis of the reflection $\alpha \circ \sigma_m \circ \alpha^{-1}$ and we have $\alpha \circ \sigma_m \circ \alpha^{-1} = \sigma_{\alpha(m)}$. The conjugation of a reflection by a translation is pictured in Figure 6.9. ∎

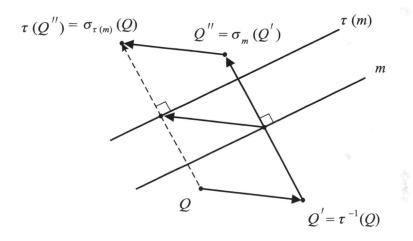

Figure 6.9. Conjugation of σ_m by τ.

Theorem 199 *Let α be an isometry, let A and B be points, let $A' = \alpha(A)$, and let $B' = \alpha(B)$.*

a. *The conjugate of a translation is a translation. In fact,*

$$\alpha \circ \tau_{\mathbf{AB}} \circ \alpha^{-1} = \tau_{\mathbf{A'B'}}.$$

b. *The conjugate of a glide reflection is a glide reflection. In fact, if $A \neq B$ and $c = \overleftrightarrow{AB}$ then*

$$\alpha \circ \gamma_{c,\mathbf{AB}} \circ \alpha^{-1} = \gamma_{\alpha(c),\mathbf{A'B'}}.$$

Proof. (a) Let M be the midpoint of $\overline{AB}$; then

$$\tau_{\mathbf{AB}} = \varphi_M \circ \varphi_A$$

by Theorem 145. Since α is an isometry, $M' = \alpha(M)$ is the midpoint of $\overline{A'B'}$. Another application of Theorem 145 gives

$$\tau_{\mathbf{A'B'}} = \varphi_{M'} \circ \varphi_{A'},$$

and by Theorem 198, part a we have

$$
\begin{aligned}
\alpha \circ \tau_{\mathbf{AB}} \circ \alpha^{-1} &= \alpha \circ (\varphi_M \circ \varphi_A) \circ \alpha^{-1} \\
&= (\alpha \circ \varphi_M \circ \alpha^{-1}) \circ (\alpha \circ \varphi_A \circ \alpha^{-1}) \\
&= \varphi_{M'} \circ \varphi_{A'} = \tau_{\mathbf{A'B'}}.
\end{aligned}
$$

The conjugation of a translation by a reflection is pictured in Figure 6.10.

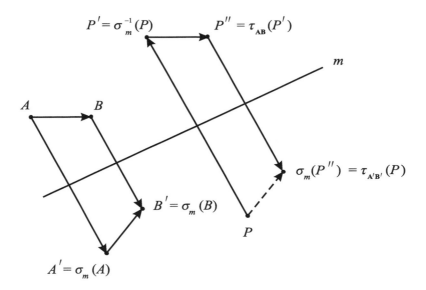

Figure 6.10. Conjugation of $\tau_{\mathbf{AB}}$ by σ_m.

(b) Since $A \neq B$ we have $\tau_{\mathbf{AB}} \neq \iota$ and $\tau_{\mathbf{AB}}(c) = c$. Hence $\gamma_{c,\mathbf{AB}} = \sigma_c \circ \tau_{\mathbf{AB}}$ by definition. Now to conjugate by α, apply Theorem 198, part b, and Theorem 199, part a, to obtain

$$\alpha \circ \gamma_{c,\mathbf{AB}} \circ \alpha^{-1} = \alpha \circ \sigma_c \circ \tau_{\mathbf{AB}} \circ \alpha^{-1} = \alpha \circ \sigma_c \circ \alpha^{-1} \circ \alpha \circ \tau_{\mathbf{AB}} \circ \alpha^{-1} = \sigma_{\alpha(c)} \circ \tau_{\mathbf{A'B'}}.$$

Since A' and B' are distinct points on line $\alpha(c)$, we have $\tau_{\mathbf{A'B'}} \neq \iota$ and $\tau_{\mathbf{A'B'}}(\alpha(c)) = \alpha(c)$. Hence $\sigma_{\alpha(c)} \circ \tau_{\mathbf{A'B'}} = \gamma_{\alpha(c),\mathbf{A'B'}}$ and it follows that $\alpha \circ \gamma_{c,\mathbf{AB}} \circ \alpha^{-1} = \gamma_{\alpha(c),\mathbf{A'B'}}$. ∎

Equation (6.5) states that the conjugate of a rotation $\rho_{O,\Theta}$ by the translation $\tau_{\mathbf{OC}}$ is the rotation $\rho_{C,\Theta}$. Since $\tau_{\mathbf{OC}}$ is even and $C = \tau_{\mathbf{OC}}(O)$, equation (6.5) is a special case of our next theorem.

Theorem 200 *The conjugate of a rotation is a rotation. In fact, if α is an isometry, C is a point, and $\Theta \in \mathbb{R}$, then*

$$\alpha \circ \rho_{C,\Theta} \circ \alpha^{-1} = \begin{cases} \rho_{\alpha(C),\Theta} & \text{if } \alpha \text{ is even,} \\ \rho_{\alpha(C),-\Theta} & \text{if } \alpha \text{ is odd.} \end{cases}$$

Proof. The Fundamental Theorem asserts that α can be expressed as a composition of three or fewer reflections. Thus we consider three cases.

<u>Case 1</u>: $\alpha = \sigma_r$. Then $\alpha = \alpha^{-1}$ is odd. If C is on r, there is a unique line s passing through C such that $\rho_{C,\Theta} = \sigma_s \circ \sigma_r$ by Corollary 142 so that

$$\sigma_r \circ \rho_{C,\Theta} \circ \sigma_r = \sigma_r \circ \sigma_s \circ \sigma_r \circ \sigma_r = \sigma_r \circ \sigma_s = \rho_{C,-\Theta} = \rho_{\sigma_r(C),-\Theta}.$$

If C is off r, let m be the line through C perpendicular to r. By Corollary 142, there is a (unique) line n passing through C such that

$$\rho_{C,\Theta} = \sigma_n \circ \sigma_m$$

(see Figure 6.11). Hence

$$\sigma_r \circ \rho_{C,\Theta} \circ \sigma_r = \sigma_r \circ \sigma_n \circ \sigma_m \circ \sigma_r = (\sigma_r \circ \sigma_n \circ \sigma_r) \circ (\sigma_r \circ \sigma_m \circ \sigma_r),$$

and by Theorem 198 we have

$$(\sigma_r \circ \sigma_n \circ \sigma_r) \circ (\sigma_r \circ \sigma_m \circ \sigma_r) = \sigma_{\sigma_r(n)} \circ \sigma_{\sigma_r(m)}. \tag{6.6}$$

Since $m \perp r$ we have

$$\sigma_r(m) = m.$$

Furthermore, $C = m \cap n$ so that

$$\sigma_r(C) = m \cap \sigma_r(n).$$

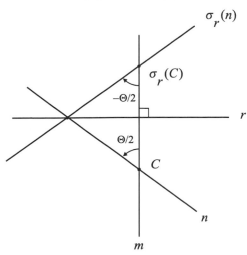

Figure 6.11. The center of rotation $\sigma_r(C) = m \cap \sigma_r(n)$.

Since the measure of an angle from m to n is $\frac{1}{2}\Theta$, the measure of an angle from m to $\sigma_r(n)$ is $-\frac{1}{2}\Theta$. By Theorem 138, the right-hand side in (6.6) becomes

$$\sigma_{\sigma_r(n)} \circ \sigma_m = \rho_{\sigma_r(C),-\Theta}$$

and we conclude that

$$\sigma_r \circ \rho_{C,\Theta} \circ \sigma_r = \rho_{\sigma_r(C),-\Theta}. \tag{6.7}$$

<u>Case 2</u>: $\alpha = \sigma_s \circ \sigma_r$. Two successive applications of (6.7) give

$$\begin{aligned}
\alpha \circ \rho_{C,\Theta} \circ \alpha^{-1} &= (\sigma_s \circ \sigma_r) \circ \rho_{C,\Theta} \circ (\sigma_s \circ \sigma_r)^{-1} \\
&= \sigma_s \circ (\sigma_r \circ \rho_{C,\Theta} \circ \sigma_r) \circ \sigma_s \\
&= \sigma_s \circ \rho_{\sigma_r(C),-\Theta} \circ \sigma_s \\
&= \rho_{\sigma_s(\sigma_r(C)),-(-\Theta)} = \rho_{\alpha(C),\Theta}.
\end{aligned}$$

<u>Case 3</u>: $\alpha = \sigma_t \circ \sigma_s \circ \sigma_r$. Three successive applications of (6.7) as in the proof of Case 2 give the desired result, as the reader can easily check. ∎

The results in Theorems 198, 199, and 200 are summarized in Theorem 201 and give us a complete picture of the action of conjugation on the set $\mathcal{I}$ of all isometries.

Theorem 201 *Let α and β be isometries. Then β and $\alpha \circ \beta \circ \alpha^{-1}$ have the same isometry type, i.e., both are reflections, both are translations, both are rotations, or both are glide reflections.*

Theorem 201 provides the critical ingredient we need to define the symmetry type of a plane figure in the next chapter. We conclude this chapter with two interesting geometrical applications of conjugation.

Theorem 202 *Two non-trivial rotations commute if and only if they have the same center.*

Proof. Let $\rho_{A,\Theta}$ and $\rho_{B,\Phi}$ be non-trivial rotations. Since $\rho_{A,\Theta}$ is even, Theorem 200 gives

$$\rho_{A,\Theta} \circ \rho_{B,\Phi} \circ \rho_{A,\Theta}^{-1} = \rho_{\rho_{A,\Theta}(B),\Phi}.$$

Hence $\rho_{B,\Phi} \circ \rho_{A,\Theta} = \rho_{A,\Theta} \circ \rho_{B,\Phi}$ if and only if $\rho_{B,\Phi} = \rho_{A,\Theta} \circ \rho_{B,\Phi} \circ \rho_{A,\Theta}^{-1} = \rho_{\rho_{A,\Theta}(B),\Phi}$ if and only if $B = \rho_{A,\Theta}(B)$ if and only if $B = A$. ∎

Theorem 203 *Two reflections σ_m and σ_n commute if and only if $m = n$ or $m \perp n$.*

Proof. Theorem 198 gives

$$\sigma_n \circ \sigma_m \circ \sigma_n = \sigma_{\sigma_n(m)}.$$

Hence $\sigma_m \circ \sigma_n = \sigma_n \circ \sigma_m$ if and only if $\sigma_m = \sigma_n \circ \sigma_m \circ \sigma_n = \sigma_{\sigma_n(m)}$ if and only if $m = \sigma_n(m)$ if and only if $m = n$ or $m \perp n$. ∎

Exercises

1. Given a line a and a point B off a, construct the line b such that $\varphi_B \circ \sigma_a \circ \varphi_B = \sigma_b$.

2. Given a line b and a point A off b, construct the point B such that $\sigma_b \circ \varphi_A \circ \sigma_b = \varphi_B$.

3. Given a line b and a point A such that $\sigma_b \circ \varphi_A \circ \sigma_b = \varphi_A$, prove that A lies on b.

4. Let A and B be distinct points and let c be a line. Prove that $\tau_{\mathbf{AB}} \circ \sigma_c = \sigma_c \circ \tau_{\mathbf{AB}}$ if and only if $\tau_{\mathbf{AB}}(c) = c$.

5. Let A and B be distinct points, let $m = \overleftrightarrow{AB}$, let $\Theta, \Phi \in \mathbb{R}$ such that $\Theta + \Phi \notin 0°$, and consider the points C and D such that $\rho_{C,\Theta+\Phi} = \rho_{B,\Phi} \circ \rho_{A,\Theta}$ and $\rho_{D,\Theta+\Phi} = \rho_{A,\Theta} \circ \rho_{B,\Phi}$. Prove that $\sigma_m(C) = D$.

6. Let a be a line and let B be a point. Prove that $\varphi_B \circ \sigma_a \circ \varphi_B \circ \sigma_a \circ \varphi_B \circ \sigma_a \circ \varphi_B$ is a reflection in some line parallel to a.

7. Let A be a point and let τ be a translation. Prove that $\varphi_A \circ \tau = \tau \circ \varphi_A$ if and only if $\tau = \iota$.

8. Prove that $\gamma_c \circ \varphi_A \neq \varphi_A \circ \gamma_c$ for every halfturn φ_A and every glide reflection γ_c with axis c.

9. Given distinct points A and B, let $\tau = \tau_{\mathbf{AB}}$ and let $\gamma_{c,\mathbf{v}}$ be a glide reflection with axis c and glide vector $\mathbf{v}$. Prove that $\tau \circ \gamma_{c,\mathbf{v}} = \gamma_{c,\mathbf{v}} \circ \tau$ if and only if $\tau(c) = c$.

10. Complete the proof of Theorem 200 by proving Case 3.

Chapter 7

Symmetry of Plane Figures

Mighty is geometry; joined with art, resistless.

Euripides (480–406 B.C.)
Greek Poet and Dramatist

Geometry is the study of those properties of a set S that remain invariant when the elements of S are subjected to the transformations of some transformation group.

Felix Klein (1849–1925)
Father of Transformational Geometry

A *symmetry* of a plane figure F is an isometry that fixes F. If F is an equilateral triangle with centroid C, for example, there are six symmetries of F, one of which is the rotation $\rho_{C,120}$. In this chapter we observe that the set of symmetries of a plane figure is a "group" with respect to composition of isometries. The structure of these groups, called *symmetry groups*, faithfully encodes the symmetries of plane figures and their interrelationships.

The Classification of Isometries (Theorem 178) asserts that the plane isometries are reflections, translations, rotations, or glide reflections. We shall use this fact to systematically identify the symmetries of a given plane figure and determine its symmetry group. Plane figures with "finitely generated" symmetry groups fall into one of five general classes: (1) *asymmetrical figures*, (2) *figures with bilateral symmetry*, (3) *rosettes with exactly one point of symmetry (the center of a non-trivial rotational symmetry)*, (4) *frieze patterns with translational symmetries in exactly one direction*, and (5) *wallpaper patterns with translational symmetries in independent directions*. There are also plane figures that are fixed by non-isometric similarities, and we shall consider them in Chapter 8. While there are infinitely many symmetry types of rosettes, it's quite surprising to find that there are exactly seven symmetry types of frieze patterns and seventeen symmetry types of wallpaper patterns.

We begin our discussion with some basic group theory.

7.1   Groups of Isometries

In this section we observe that the set $\mathcal{I}$ of all isometries is a group with respect to composition, then consider some of its subgroups. We also prove the somewhat surprising fact that whenever a finite group of isometries contains reflections, the number of reflections always equals the number of rotations (including the identity).

Definition 204 *Let G be a non-empty set. A **binary operation** on G is a function $* : G \times G \to G$ with domain $G \times G$. If $a, b \in G$, the symbol $a * b$ denotes the image of (a, b) under the operation $*$. If H is a non-empty subset of G and a binary operation $*$ on G restricts to a binary operation on H, we say that $*$ is **closed in** H.*

Definition 205 *Let G be a non-empty set equipped with a binary operation $*$. Then $(G, *)$ is a **group** if the following axioms are satisfied:*

1. *Associativity: If $a, b, c \in G$, then $(a * b) * c = a * (b * c)$.*

2. *Two-sided identity: For all $a \in G$, there exists an element $e \in G$ such that $e * a = a * e = a$.*

3. *Existence of inverses: For each $a \in G$, there exists $b \in G$ such that $a * b = b * a = e$. The element b is called the **inverse of** a and we write $b = a^{-1}$.*

*A group $(G, *)$ is **abelian** (or **commutative**) if for all $a, b \in G$, $a*b = b*a$.*

In fact, a group G has a *unique* two-sided identity and the inverse of each element is *unique*. We leave the proofs of uniqueness as exercises for the reader.

Theorem 206 *The set $\mathcal{I}$ of all isometries is a group with respect to composition.*

Proof. The work has already been done. Composition is a binary operation since the composition of an isometry is an isometry by Exercise 3.1.2, and associativity is a special case of Exercise 3.1.4. The fact that the identity transformation ι acts as a two-sided identity element was proved in Exercise 3.1.3, and the existence of inverses was proved in Exercise 3.1.5. ∎

Since two halfturns with distinct centers do not commute, the group $(\mathcal{I}, \circ)$ is *non-abelian*.

Definition 207 *A non-empty subset H of a group $(G, *)$ is a **subgroup** of G if $*$ is closed in H and $(H, *)$ is a group.*

Some subsets of $\mathcal{I}$ contain elements that commute with each other. Two such examples are (1) the subset $\mathcal{T}$ of all translations and (2) the subset $\mathcal{R}_C$ of all rotations about a point C. The fact that $\mathcal{T}$ and $\mathcal{R}_C$ are *abelian groups* is easy to verify by applying our next theorem.

Theorem 208 *A non-empty subset H of a group $(G, *)$ is a subgroup of G if and only if the following axioms are satisfied:*

1. *Closure: The operation $*$ on G is closed in H.*

2. *Existence of inverses: If $a \in H$, then $a^{-1} \in H$.*

Proof. ($\Rightarrow$) If H is a subgroup of G, axioms (1) and (2) hold by definition.

($\Leftarrow$) Let H be a non-empty subset of G satisfying axioms (1) and (2). *Associativity:* If $a, b, c \in H$, then $(a * b) * c \in H$ and $a * (b * c) \in H$ by axiom (1). But $(a * b) * c = a * (b * c)$ as elements of G. *Two-sided identity:* Since $H \neq \varnothing$, choose any element $a \in H$. Then $a^{-1} \in H$ by axiom (2) and $a * a^{-1} \in H$ by axiom (1). But $a * a^{-1} = e$, the identity element in G, so that $e \in H$. Therefore H is a subgroup of G. ■

Proposition 209 *The set $\mathcal{T}$ of all translations is an abelian group.*

Proof. Closure, the existence of inverses, and commutativity were proved in Proposition 106. Therefore $\mathcal{T}$ is an abelian subgroup of $\mathcal{I}$ by Theorem 208. ■

Proposition 210 *The set $\mathcal{R}$ of all rotations about the origin is an abelian group.*

Proof. The proof is left as an exercise for the reader. ■

Notation 211 *Let $(G, *)$ be a group and let $a, b \in G$. We sometimes use juxtaposition ab to denote $a * b$. Define $a^0 := e$; for $n \in \mathbb{N}$ define*

$$a^n := \underbrace{aa \cdots a}_{n \text{ factors}} \quad and \quad a^{-n} := \left(a^{-1}\right)^n.$$

Definition 212 *A group G is **cyclic** if there is an element $g \in G$, called a **generator**, such that $G = \left\{g^k \mid k \in \mathbb{Z}\right\}$. A finite cyclic group G with n elements is **cyclic of order** n. A cyclic group G with infinitely many elements is **infinite cyclic**.*

Cyclic groups are always abelian. To see this, let $(G, *)$ be a cyclic group and choose a generator $g \in G$. If $a, b \in G$, there exit $m, n \in \mathbb{Z}$ such that $a = g^m$ and $b = g^n$. Hence $ab = g^m g^n = g^{m+n} = g^{n+m} = g^n g^m = ba$.

Theorem 213 *Every finite group of isometries G has the following properties:*

a. *The elements of G are rotations or reflections.*

b. *All non-trivial rotations in G have the same center.*

c. *The set of all rotations in G forms a cyclic subgroup.*

d. *If G has reflections, the number of reflections equals the number of rotations.*

Proof. (a) Since G is finite, and non-trivial translations and glide reflections generate infinite subgroups, the elements of G are rotations or reflections.

(b) Suppose $\rho_{C,\Theta}, \rho_{B,\Phi} \in G$ are non-trivial rotations with $B \neq C$. Then $C' = \rho_{B,\Phi}(C) \neq C$ since B is the center of rotation, and $\rho_{B,\Phi} \circ \rho_{C,\Theta} \circ \rho_{B,\Phi}^{-1} = \rho_{C',\Theta}$ by conjugation. By closure, $\rho_{C',\Theta} \in G$ and $\rho_{C',\Theta} \circ \rho_{C,\Theta}^{-1} = \rho_{C',\Theta} \circ \rho_{C,-\Theta} \in G$. But by Theorem 157 (the Angle Addition Theorem), part 2, $\rho_{C',\Theta} \circ \rho_{C,-\Theta}$ is a non-trivial translation since the sum of the rotation angles $\Theta + (-\Theta) \in 0°$, which is a contradiction. Therefore $B = C$.

(c) Let E be the set of all rotations in G, and write each element of E uniquely in the form ρ_Φ with $0 \leq \Phi < 360$. Since E is finite, there is a rotation ρ_Θ with the smallest positive rotation angle Θ. Let $\rho_{C,\Phi} \in E$ be a non-trivial rotation. Then $\Theta \leq \Phi < 360$ by the minimality of Θ, and there is some integer $k \geq 1$ such that $k\Theta \leq \Phi < (k+1)\Theta$, or equivalently, $0 \leq \Phi - k\Theta < \Theta$. Note that $\rho_{\Phi - k\Theta} = \rho_\Phi \circ \rho_\Theta^{-k} \in G$ by closure, so if $0 < \Phi - k\Theta < \Theta$, the minimality of Θ is violated. Hence $\Phi - k\Theta = 0$ so that $\Phi = k\Theta$. We conclude that $\rho_{C,\Phi} = \rho_{C,k\Theta} = \rho_{C,\Theta}^k$. Therefore $E = \left\{ \rho_\Theta, \rho_\Theta^2, \ldots, \rho_\Theta^n = \iota \right\}$ for some $n \geq 1$ and E is a cyclic subgroup of order n.

(d) Let $F = \{\sigma_1, \sigma_2, \ldots, \sigma_m\}$ be the set of all reflections in G. The elements of $\sigma E = \left\{ \sigma \circ \rho_\Theta, \sigma \circ \rho_\Theta^2, \ldots, \sigma \circ \rho_\Theta^n \right\}$ are reflections since they have odd parity and G contains no glide reflections. These reflections are all distinct since $\sigma \circ \rho_\Theta^i = \sigma \circ \rho_\Theta^j$ with $i, j \leq n$ implies $i = j$. Hence $\sigma E \subseteq F$ and $n \leq m$. Similarly, the elements of $\sigma F = \{\sigma \circ \sigma_1, \sigma \circ \sigma_2, \ldots, \sigma \circ \sigma_m\}$ are rotations since they have even parity and G contains no translations. These rotations are all distinct since $\sigma \circ \sigma_i = \sigma \circ \sigma_j$ implies $i = j$. Hence $\sigma F \subseteq E$ and $m \leq n$. Therefore $m = n$. ∎

Exercises

1. Let G be a group. Prove that there is a unique identity element $e \in G$.

2. Let G be a group. Prove that an element $g \in G$ has a unique inverse.

3. Prove that the set $\mathcal{R}$ of all rotations about the origin is an abelian group.

4. Prove that the set $\mathcal{E}$ of all even isometries is a non-abelian group.

5. Prove that the set $\mathcal{D}$ of all isometric dilatations is a non-abelian group.

7.2 Symmetry Type

In this section we observe that the set of symmetries of a given plane figure F is a group, called the *symmetry group* of F. Of course, every symmetry group is a subgroup of the group $\mathcal{I}$ of all isometries. We introduce an equivalence relation on the family of all symmetry groups, called *inner isomorphism*, and define the *symmetry type* of a plane figure to be the inner isomorphism class of its symmetry group.

Inner isomorphisms are defined by conjugation, and conjugation preserves isometry type. Consequently, in addition to having the same group structures, symmetry groups of plane figures representing the same inner isomorphism class contain the same symmetries. Thus inner isomorphism classes of symmetry groups of plane figures provide a perfect algebraic invariant in the sense that two planes figures have the same symmetry type if and only if their symmetry groups are inner isomorphic.

Definition 214 *A **symmetry** of a plane figure F is an isometry that fixes F. A plane figure with non-trivial rotational symmetry has **point symmetry** and the center of rotation is called a **point of symmetry**. A plane figure with reflection symmetry has **line symmetry** and the axis of reflection is called a **line of symmetry**. A plane figure with **bilateral symmetry** has a unique line of symmetry.*

Theorem 215 *Given a plane figure F, the set $\mathrm{Sym}\,(F)$ of all symmetries of F is a group, called the **symmetry group** of F.*

Proof. Note that $\mathrm{Sym}\,(F) \neq \varnothing$ since $\iota \in \mathrm{Sym}\,(F)$. Since $\mathrm{Sym}\,(F)$ is a subset of the group $\mathcal{I}$ of all isometries, it suffices to verify the axioms of Theorem 208.

<u>Closure:</u> Given $\alpha, \beta \in \mathrm{Sym}\,(F)$, we have $\alpha\,(F) = F$ and $\beta\,(F) = F$. Since the composition of isometries is an isometry, by Exercise 3.1.2, and $(\alpha \circ \beta)\,(F) = \alpha\,(\beta\,(F)) = \alpha\,(F) = F$. Therefore $\alpha \circ \beta \in \mathrm{Sym}\,(F)$.

<u>Existence of inverses:</u> If $\alpha \in \mathrm{Sym}\,(F)$, then $\alpha\,(F) = F$ and $\alpha^{-1} \in \mathcal{I}$ by Exercise 3.1.5. But $\alpha^{-1}\,(F) = \alpha^{-1}\,(\alpha\,(F)) = (\alpha^{-1} \circ \alpha)\,(F) = \iota\,(F) = F$ so that $\alpha^{-1} \in \mathrm{Sym}\,(F)$. ∎

Example 216 (The Dihedral Group D_3) Let T denote an equilateral triangle positioned with its centroid at the origin and a vertex on the y-axis as in Figure 7.1. There are exactly six symmetries of T, namely, the identity ι, two non-trivial rotations ρ_{120} and ρ_{240} about the origin, and three reflections σ_l, σ_m, and σ_n, where $l : \sqrt{3}X - 3Y = 0$, $m : X = 0$, and $n : \sqrt{3}X + 3Y = 0$.

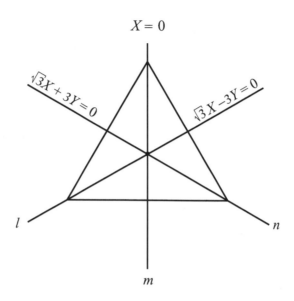

Figure 7.1. Lines of symmetry l, m, and n; the origin is a point of symmetry.

The Cayley table (multiplication table) for the various compositions of these symmetries appears in Table 7.1. By Theorem 215 these six symmetries form a group, which we denote by D_3. The symbol D_3 stands for the *"Dihedral Group of Order 6."* Thus $\text{Sym}(T) = D_3$.

Example 217 (The Cyclic Group C_3) The upper left 4×4 block in Table 7.1 is the Cayley table for the subset $C_3 = \{\iota, \rho_{120}, \rho_{240}\}$ of rotations in D_3 (the identity is a trivial rotation). By inspection we see that composition is closed in C_3 and each element of C_3 has an inverse in C_3. Therefore C_3 is a group by Theorem 215. Furthermore, note that the Cayley table for C_3 is a symmetric matrix. This visually indicates that C_3 is abelian. The symbol C_3 stands for the *"Cyclic Group of Order 3."*

The polygons in the examples that follow are momentarily positioned with their centroids at the origin and a line of symmetry along the y-axis as in Figure 7.1. However, we will drop this requirement shortly.

$\circ$	ι	ρ_{120}	ρ_{240}	σ_l	σ_m	σ_n
ι	ι	ρ_{120}	ρ_{240}	σ_l	σ_m	σ_n
ρ_{120}	ρ_{120}	ρ_{240}	ι	σ_m	σ_n	σ_l
ρ_{240}	ρ_{240}	ι	ρ_{120}	σ_n	σ_l	σ_m
σ_l	σ_l	σ_n	σ_m	ι	ρ_{240}	ρ_{120}
σ_m	σ_m	σ_l	σ_n	ρ_{120}	ι	ρ_{240}
σ_n	σ_n	σ_m	σ_l	ρ_{240}	ρ_{120}	ι

TABLE 7.1: The Cayley Table for the Dihedral Group D_3 of Order 6.

Example 218 (The Dihedral Group D_n) A 90-120-kite has one line of (bilateral) symmetry and one rotational symmetry (the identity). These two symmetries form the Dihedral Group of Order 2, denoted by D_1. A non-square rhombus has two lines of symmetry – its diagonals – and two rotational symmetries – the identity and the halfturn about the origin. These four symmetries form the Dihedral Group of Order 4, denoted by D_2. A regular n-gon, $n \geq 3$, has n lines of symmetry – the perpendicular bisectors of its sides and the bisectors of its interior angles – and n rotational symmetries of $\left(\frac{360}{n}k\right)^\circ$, $0 \leq k \leq n-1$. The $2n$ symmetries of a regular n-gon form the Dihedral Group of Order $2n$, denoted by D_n. Note that if a polygon P has symmetry group D_n, its lines of symmetry pass through the origin. Thus when $n \geq 2$, all lines of symmetry are concurrent at the origin.

Example 219 (The Cyclic Group C_n) Let ρ_Θ denote the rotation about the origin of Θ°. Observe that the elements of $C_3 = \{\iota, \rho_{120}, \rho_{240}\}$ are powers of either ρ_{120} or ρ_{240}. For example,

$$\rho_{120} = \rho_{120}^1; \quad \rho_{240} = \rho_{120}^2; \text{ and } \iota = \rho_{120} \circ \rho_{240} = \rho_{120}^3.$$

Thus ρ_{120} or ρ_{240} generates C_3. More generally, let n be a positive integer and let $\Theta = \frac{360}{n}$. Then $\rho_\Theta^k = \rho_{k\Theta}$ for each integer k, and in particular, $\rho_\Theta^n = \rho_{360} = \iota$. Thus Θ gives rise to exactly n congruence classes of rotation angles $\Theta^\circ, 2\Theta^\circ, \ldots, n\Theta^\circ$, and the cyclic group generated by ρ_Θ has exactly n elements. Thus $C_n = \{\rho_\Theta, \rho_\Theta^2, \ldots, \rho_\Theta^n = \iota\}$.

For $n \geq 3$, the cyclic group C_n discussed in Example 219 is the symmetry group of a (non-regular) $3n$-gon constructed from a regular n-gon in the following way: For $n = 4$, cut a square out of paper and draw its diagonals, thereby subdividing the square into four congruent isosceles right triangles with common vertex at the centroid of the square. From each of the four vertices, cut along the diagonals stopping midway between the vertices and the centroid. With the square positioned so that its edges are vertical or horizontal, fold the triangle at the top so that its right-hand vertex aligns with the centroid of the square. Rotate the paper 90° and fold the triangle now at the top in a similar way. Continue rotating and folding until you have what looks like a flattened pinwheel with four blades (see Figure 7.2). The outline of this flattened pinwheel is a dodecagon (12-gon) with symmetry group C_4 generated by either ρ_{90} or ρ_{270}. In a similar way, one can cut and fold a regular n-gon to obtain pinwheel with n blades and symmetry group C_n.

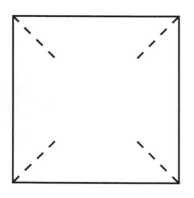   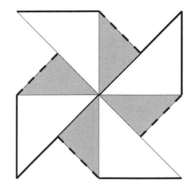

Cut along the dotted lines   Fold along the dotted lines

Figure 7.2. A polygon with symmetry group C_4.

Example 220 Let τ be a non-trivial translation, let P be any point, and let $Q = \tau(P)$. Then $\tau = \tau_{PQ}$ and $\tau^2 = \tau_{PQ} \circ \tau_{PQ} = \tau_{2PQ}$. Inductively, $\tau^n = \tau^{n-1} \circ \tau = \tau_{(n-1)PQ} \circ \tau_{PQ} = \tau_{nPQ}$, for each $n \in \mathbb{N}$. Furthermore, $\tau^0 = \iota$ and $\tau^{-n} = (\tau^{-1})^n = \tau^n_{-PQ} = \tau_{-nPQ}$ for each $n \in \mathbb{N}$. Hence distinct integer powers of τ are distinct translations and it follows that $G = \{\tau^n \mid n \in \mathbb{Z}\}$ is an infinite set. To see that G is an infinite cyclic group generated by τ, we first apply Theorem 208 to check that G is a subgroup.

<u>Closure:</u> For all $m, n \in \mathbb{Z}$, $\tau^n \circ \tau^m = \tau_{nPQ} \circ \tau_{mPQ} = \tau_{(n+m)PQ} = \tau^{n+m}$.

<u>Existence of inverses:</u> For all $n \in \mathbb{Z}$, $\tau^{-n} \circ \tau^n = \tau_{-nPQ} \circ \tau_{nPQ} = \iota$. Therefore $(\tau^n)^{-1} = \tau^{-n}$.

Finally, since every element of G is an integer power of τ (or τ^{-1}), G is infinite cyclic and generated by τ (or τ^{-1}).

Let G be a group and let K be a non-empty subset of G. The symbol $\langle K \rangle$ denotes the set of all (finite) products of powers of elements of K and their inverses. If $K = \{k_1, k_2, \ldots\}$, we abbreviate and write $\langle k_1, k_2, \ldots \rangle$ instead of $\langle \{k_1, k_2, \ldots\} \rangle$. Thus $\langle K \rangle$ is automatically a subgroup of G since by definition $\langle K \rangle$ is non-empty, the group operation is closed in $\langle K \rangle$, and $\langle K \rangle$ contains the inverse of each of its elements. When G is cyclic with generator g we have $G = \langle g \rangle$.

Definition 221 *Let G be a group and let K be a non-empty subset of G. The subgroup $\langle K \rangle$ is called the **subgroup of** G **generated by** K. A subset $K \subseteq G$ is a **generating set** for G if $G = \langle K \rangle$. A group G is **finitely generated** if there exists a finite generating set K for G.*

Example 222 Consider a rotation ρ about the origin O of $\left(\frac{360}{n}\right)^{\circ}$ and a reflection σ in the y-axis. Then $\rho, \sigma \in D_n$ and $\langle \rho, \sigma \rangle$ is a subgroup of D_n. On the other hand, the subgroup of rotations in D_n is $C_n = \left\{\rho, \rho^2, \ldots, \rho^n\right\}$, and the elements of $\sigma C_n = \left\{\sigma \circ \rho, \sigma \circ \rho^2, \ldots, \sigma \circ \rho^n\right\}$ are reflections, since they are odd and not glide reflections (the square of a glide reflection is a non-trivial translation). Furthermore, the elements of σC_n are distinct, since $\sigma \circ \rho^j = \sigma \circ \rho^k$ with $1 \leq j, k \leq n$ implies $j = k$, and C is on the axis of $\sigma \circ \rho^i$ since $\left(\sigma \circ \rho^i\right)(C) = \sigma\left(C\right) = C$. Therefore

$$D_n = C_n \cup \sigma C_n = \left\{\rho, \rho^2, \ldots, \rho^n\right\} \cup \left\{\sigma \circ \rho, \sigma \circ \rho^2, \ldots, \sigma \circ \rho^n\right\} \subseteq \langle \rho, \sigma \rangle$$

and it follows that $D_n = \langle \rho, \sigma \rangle$.

Example 223 Let F be the set of all reflections. Since every reflection is its own inverse, $\langle F \rangle$ consists of all (finite) compositions of reflections. By the Fundamental Theorem every isometry is a composition of three or fewer reflections. Therefore the group of all isometries $\mathcal{I} = \langle F \rangle$. To see that $\mathcal{I}$ is infinitely generated, note that a finite set of reflections generates at most a countable number of reflections. Since the number of reflections in $\mathcal{I}$ is uncountable, no finite number of reflections generates $\mathcal{I}$.

Example 224 Consider the set H of all halfturns. Since the composition of two halfturns is a translation, composition is not closed in H, and H is not a group. However, the composition of two translations is a translation and the composition of a translation and a halfturn is a halfturn. Hence $\langle H \rangle$ is exactly the group of all translations and halfturns (the isometric dilatations).

So far, we have limited our discussion of symmetry to polygons with line symmetry along the y-axis and point symmetry at the origin, and as we have seen, the symmetry groups of these polygons are either D_n or C_n. But what is the symmetry group G of a regular n-gon whose centroid is at some point A off the y-axis? Clearly, $G \neq D_n$. Nevertheless, G and D_n are in some sense identical, and we would like to identify them appropriately. This is the goal of the discussion that follows.

Definition 225 *Let $(G, \cdot)$ and $(G', *)$ be groups. A **homomorphism** from G to G' is a function $\phi : G \to G'$ that preserves group operations, i.e., if $a, b \in G$, then*

$$\phi\left(a \cdot b\right) = \phi\left(a\right) * \phi\left(b\right).$$

*An **isomorphism** from G to G' is a bijective homomorphism $\phi : G \to G'$. When an isomorphism from G to G' exists, we say that G and G' are **isomorphic** and write $G \approx G'$. An **automorphism** of G is an isomorphism $\phi : G \to G$.*

Let $e \in G$ and $e' \in G'$ be identity elements, let $g \in G$, and let $\phi : G \to G'$ be a homomorphism. It is easy to check that $\phi(e) = e'$ and $\phi(g^{-1}) = [\phi(g)]^{-1}$. Thus ϕ is completely determined by its action on a generating set. For example, if $G = \langle g, h \rangle$, the action of ϕ on the product $g \cdot h$ is given by $\phi(g \cdot h) = \phi(g) * \phi(h)$. This reminds us of the fact that a linear map of vector spaces is completely determined by its action on a basis.

Example 226 Let G be a group and let $a \in G$. Define a function $\phi_a : G \to G$ by $\phi(x) = axa^{-1}$. Then ϕ_a is a homomorphism since for all $g, h \in G$ we have

$$\phi_a(gh) = agha^{-1} = \left(aga^{-1}\right)\left(aha^{-1}\right) = \phi_a(g)\,\phi_a(h).$$

Furthermore, if $\phi_a(g) = \phi_a(h)$, then $aga^{-1} = aha^{-1}$ and $g = h$ so that ϕ_a is injective, and if $g \in G$, then $\phi_a\left(a^{-1}ga\right) = a\left(a^{-1}ga\right)a^{-1} = g$ so that ϕ_a is surjective. Hence ϕ_a is an automorphism of G.

The automorphism ϕ_a in Example 226 is quite special because it acts by conjugation. Since conjugation preserves isometry type (Theorem 201), automorphisms ϕ_α of $\mathcal{I}$ are fundamentally important and need to be distinguished from automorphisms of $\mathcal{I}$ that cannot be defined by conjugation.

Definition 227 *Let G be a group and let $a \in G$. The automorphism $\phi_a : G \to G$ defined by $\phi_a(x) = axa^{-1}$ is called an **inner automorphism** of G. Automorphisms of G that cannot be defined by conjugation are called **outer automorphisms**. Two subgroups $H, H' \subseteq G$ are **inner isomorphic** if there exists an inner automorphism $\phi_a : G \to G$ such that $\phi_a(H) = H'$, in which case we write $H \approx_I H'$ and refer to the restriction of ϕ_a to H as an **inner isomorphism** from H to H'.*

Example 228 Consider the symmetry group G of a regular n-gon with centroid at some point $A \neq O$. Let us construct an inner isomorphism $\phi_\alpha : G \to D_n$. Choose a reflection $\sigma_c \in G$ and let $\Theta = \frac{360}{n}$; then $G = \langle \rho_{A,\Theta}, \sigma_c \rangle$ and the axis of reflection c is either parallel to or cuts the y-axis m. Let $\tau = \tau_{\mathbf{AO}}$.

<u>Case 1</u>: $c \parallel m$. Since A is on c, O is on m, and τ is a dilatation we have $\tau(c) = m$. Consider the inner automorphism $\phi_\tau : \mathcal{I} \to \mathcal{I}$ defined by $\phi_\tau(x) = \tau \circ x \circ \tau^{-1}$. Then on generators we have

$$\phi_\tau(\rho_{A,\Theta}) = \tau \circ \rho_{A,\Theta} \circ \tau^{-1} = \rho_{\tau(A),\Theta} = \rho_{O,\Theta} \text{ and}$$

$$\phi_\tau(\sigma_c) = \tau \circ \sigma_c \circ \tau^{-1} = \sigma_{\tau(c)} = \sigma_m.$$

Thus ϕ_τ restricts to an inner isomorphism $\phi_\tau : G \to D_n$ and $G \approx_I D_n$.

<u>Case 2</u>: c cut m. Let Φ be the measure of an angle from c to m and let $\alpha = \tau \circ \rho_{A,\Phi}$. Then $\alpha(A) = (\tau \circ \rho_{A,\Phi})(A) = \tau(A) = O$ and $c' = \rho_{A,\Phi}(c) \parallel m$. Since

A is on c', O is on m, and τ is a dilatation we have $\alpha\left(c\right) = \left(\tau \circ \rho_{A,\Phi}\right)\left(c\right) = \tau\left(c'\right) = m$. Consider the inner automorphism $\psi_\alpha : \mathcal{I} \to \mathcal{I}$ defined by $\psi_\alpha\left(x\right) = \alpha \circ x \circ \alpha^{-1}$. Then on generators we have

$$\psi_\alpha\left(\rho_{A,\Theta}\right) = \alpha \circ \rho_{A,\Theta} \circ \alpha^{-1} = \rho_{\alpha(A),\Theta} = \rho_{O,\Theta} \text{ and}$$

$$\psi_\alpha\left(\sigma_c\right) = \alpha \circ \sigma_c \circ \alpha^{-1} = \sigma_{\alpha(c)} = \sigma_m.$$

Thus ψ_α restricts to an inner isomorphism $\psi_\alpha : G \to D_n$ and $G \approx_I D_n$.

The inner isomorphisms ϕ_τ and ψ_α in Example 228 preserve the "symmetry type" of the underlying plane figures. Let's make this idea precise. If F_1 and F_2 are *congruent* plane figures with respective symmetry groups G_1 and G_2, there is an isometry α such that $F_2 = \alpha\left(F_1\right)$ and $G_2 = \phi_\alpha\left(G_1\right)$. On the other hand, an inner isomorphism $\phi_\alpha : G_1 \to G_2$ exists whenever F_1 and F_2 have the same symmetries – whether or not they are congruent. For example, although a snowflake and a regular hexagon are not congruent, they have exactly the same symmetries and there is an inner isomorphism of symmetry groups defined as in Example 228. Thus *symmetry groups record the symmetries of a plane figure and not its precise shape.* Furthermore, *inner isomorphism* is an equivalence relation on the family of all symmetry groups, and the inner isomorphism classes are the symmetry types of plane figures.

Definition 229 *The **symmetry type** of a plane figure F with symmetry group G is the inner isomorphism class*

$$[G] = \{\phi_\alpha\left(G\right) \mid \alpha \in \mathcal{I}\}.$$

Two plane figures with inner isomorphic symmetry groups have the same symmetry type.

It is now clear that the inner isomorphism class $[D_n]$ contains the symmetry groups of all plane figures with the same symmetries as some regular n-gon and all such plane figures have the same symmetry type. Henceforth, *the name assigned to a particular symmetry group will also be assigned to every symmetry group in its inner isomorphism class.* Thus every plane figure with the same symmetries as some regular n-gon has symmetry group D_n.

We conclude this section with an example demonstrating the fact that inner isomorphisms of symmetry groups are more discriminating than usual group isomorphisms.

Example 230 The symmetry groups $D_1 = \{\iota, \sigma\}$ of a butterfly and $C_2 = \{\iota, \varphi\}$ of a yin-yang symbol are isomorphic but not inner isomorphic since inner automorphisms of $\mathcal{I}$ never send reflections to halfturns. Thus the symmetry types of a butterfly and yin-yang symbol are distinct.

Exercises

1. Recall that the six symmetries of an equilateral triangle form the dihedral group D_3 (see Example 216). Show that the set $K = \{\rho_{120},\ \sigma_l\}$ is a generating set for D_3 by writing each of the other four elements in D_3 as a product of powers of elements of K and their inverses. Compute all powers of each element in D_3 and show that no single element alone generates D_3. Thus D_3 is not cyclic.

2. The dihedral group D_4 consists of the eight symmetries of a square. When a square is positioned with its vertices on the axes, the origin is a point of symmetry and the lines $a : Y = 0$, $b : Y = X$, $c : X = 0$ and $d : Y = -X$ are lines of symmetry. Construct the Cayley table for $D_4 = \{\iota, \rho_{90}, \rho_{180}, \rho_{270}, \sigma_a, \sigma_b, \sigma_c, \sigma_d\}$.

3. Find the symmetry group of the following:

   (a) A parallelogram that is neither a rectangle nor a rhombus.
   (b) A rectangle that is not a square.

4. Find the symmetry group of the capital letter "O" written as a non-circular oval. Find the symmetry group of the other 25 capital letters of the alphabet written in most symmetric form.

5. Determine the symmetry group of each figure below:

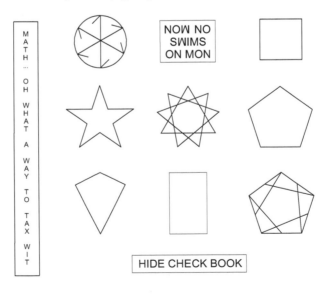

6. The discussion following Example 219 describes how to construct a $3n$-gon whose symmetry group is C_n, where $n \geq 3$. Alter this construction to obtain a $2n$-gon whose symmetry group is C_n.

7. Prove that a plane figure with bilateral symmetry has no points of symmetry.

8. Let C be a point and let $\Theta \in \mathbb{R}$. Prove that $\langle \rho_{C,\Theta} \rangle$ is finite if and only if $\Theta \in \mathbb{Q}$.

9. Consider the dihedral group D_n.

    (a) Prove that D_n contains a halfturn φ if and only if n is even.
    (b) Prove that if D_n contains a halfturn φ, then $\varphi \circ \alpha = \alpha \circ \varphi$ for all $\alpha \in D_n$.
    (c) Let $n \geq 3$ and let $\beta \in D_n$ such that $\beta \neq \iota$. Prove that if $\beta \circ \alpha = \alpha \circ \beta$ for all $\alpha \in D_n$, then β is a halfturn. (The subgroup of elements that commute with every element of a group G is called the *center of G*. Thus, if D_n contains a halfturn φ, the center of D_n is the subgroup $\{\iota, \varphi\}$.)

10. Prove that if $\phi : \mathcal{I} \to \mathcal{I}$ is an inner automorphism and G is a subgroup of $\mathcal{I}$, then $\phi(G)$ is a subgroup of $\mathcal{I}$.

11. Let G and G' be groups, let $e \in G$ and $e' \in G'$ be identity elements, let $g \in G$, and let $\phi : G \to G'$ be a homomorphism. Prove that

    (a) $\phi(e) = e'$.
    (b) $\phi\left(g^{-1}\right) = [\phi(g)]^{-1}$.

12. Prove that inner isomorphism is an equivalence relation on the set of all symmetry groups.

7.3   Rosettes

Typically one thinks of a rosette as a pinwheel as in Figure 7.2 or a flower with n-petals as in Figure 7.3. However, a regular polygon, a non-square rectangle, a yin-yang symbol, and a pair of perpendicular lines are rosettes as well. Rosettes are in some sense the simplest plane figures with non-trivial symmetries. The symmetry group of a rosette is either D_n or C_n for some $n \geq 2$. Thus for each $n \geq 2$, there are exactly two symmetry types of rosettes.

Figure 7.3. A typical rosette.

Recall that each real number is congruent (mod 360) to exactly one real number in the interval $[0, 360)$. Thus for simplicity of exposition, all rotation angles in this section are assumed to lie in $[0, 360)$ unless explicitly indicated otherwise.

Definition 231 *Let F be a plane figure. A non-trivial rotation $\rho_{C,\Theta} \in$ Sym(F) is **minimal** if $\Theta \leq \Phi$ for every non-trivial rotation $\rho_{D,\Phi} \in$ Sym(F).*

Definition 232 *A **rosette** is a plane figure R with the following properties:*

1. *There is a minimal rotational symmetry of R.*

2. *All non-trivial rotational symmetries of R have the same center.*

*The symmetry group of a rosette is called a **rosette group**.*

Theorem 233 *A rosette group G has the following properties:*

a. *G contains only rotations and reflections.*

b. *The set of all rotations in G forms a finite cyclic subgroup.*

c. *G contains at most finitely many reflections.*

Proof. Let $\rho_\Theta = \rho_{C,\Theta}$ be the minimal rotation in G.

(a) Suppose G contains a non-trivial translation τ. Then $\rho_{\tau(C),\Theta} = \tau \circ \rho_{C,\Theta} \circ \tau^{-1} \in G$ by closure and $\tau(C) \neq C$. But all rotations in G have the same center, so this is a contradiction. Furthermore, since the square of a glide reflection is a non-trivial translation by Theorem 168, part c, G contains no glide reflections.

(b) Let $\rho_\Phi \in G$ be a non-trivial rotation. Then $\Theta \leq \Phi$ by the minimality of Θ, and there is a positive integer m such that $m\Theta \leq \Phi < (m+1)\Theta$, or equivalently, $0 \leq \Phi - m\Theta < \Theta$. Note that $\rho_{\Phi - m\Theta} = \rho_\Phi \circ \rho_{-m\Theta} = \rho_\Phi \circ \rho_{m\Theta}^{-1} \in G$ by closure. If $0 < \Phi - m\Theta < \Theta$, then $\rho_{\Phi - m\Theta}$ is a rotational symmetry with $0 < \Phi - m\Theta < \Theta$, which is a contradiction. Therefore $\Phi - m\Theta = 0$ so that $\Phi = m\Theta$ and we have $\rho_\Phi = \rho_{m\Theta} = \rho_\Theta^m$. Now suppose $\langle \rho_\Theta \rangle$ is infinite. Then $\rho_\Theta^n = \rho_{n\Theta} \neq \iota$ and $n\Theta \not\equiv 0°$ for all $n \geq 1$. Since $0 < \Theta < 360$, there is a positive integer $m \geq 1$ such that $m\Theta < 360 < (m+1)\Theta$. Hence $0 < 360 - m\Theta < \Theta$ and $\rho_{360 - m\Theta} = \rho_{-m\Theta} = \rho_{m\Theta}^{-1}$ is a rotational symmetry in G with $0 < 360 - m\Theta < \Theta$, which is a contradiction.

(c) If G has no reflections, there is nothing to prove. So assume G contains a reflection σ. Let F be the set of all reflections in G and consider the set $\sigma F = \{ \sigma \circ \sigma' \mid \sigma' \in F \}$. Define a function $f : F \to \sigma F$ by $f(\sigma') = \sigma \circ \sigma'$. We claim that f is bijective. For injectivity, assume that $f(\sigma') = f(\sigma'')$. Then $\sigma \circ \sigma' = \sigma \circ \sigma''$ implies $\sigma' = \sigma''$. For surjectivity, let $y \in \sigma F$. Then by definition of σF, there is a reflection $\sigma' \in F$ such that $y = \sigma \circ \sigma'$, and $f(\sigma') = \sigma \circ \sigma' = y$. Thus F and σF have the same cardinality. Furthermore, the elements of σF are rotations since their parity is even and G has no translations. Since G has finitely many rotations, σF is finite and so is F. ∎

In the early sixteenth century, the famous artist Leonardo da Vinci (1452–1519) determined all possible finite groups of isometries, all but two of which are rosette groups. The two exceptions are C_1, which contains only the identity, and D_1, which contains the identity and one reflection.

Theorem 234 (Leonardo Da Vinci) *Every finite group of isometries is either C_n or D_n for some $n \geq 1$.*

Proof. Let G be a finite group of isometries. Then the elements of G are rotations or reflections by Theorem 213, part a. If G has only the trivial rotation, it has at most one reflection σ and either $G = C_1 = \{\iota\}$ or $G = D_1 = \{\iota, \sigma\}$. So assume G has non-trivial rotations. Since G is finite, it has a minimal rotation ρ_Θ. By Theorem 213, part c, every rotation in G is an element of the cyclic subgroup $\langle \rho_\Theta \rangle$, which has order $n \geq 2$. If G contains no reflections, then $G = C_n$. If G contains reflections, it contains exactly n reflections $\sigma_1, \sigma_2, \ldots, \sigma_n$ by Theorem 213, part d. But all non-trivial rotations in G have the same center by Theorem 213, part b, and $\sigma_j \circ \sigma_i$ is a non-trivial rotation whenever $i \neq j$. Thus all axes of reflection are concurrent at the center of rotation and $G = D_n$. ∎

Since rosette groups are finite, the classification of rosettes up to symmetry type is an immediate consequence of Da Vinci's Theorem.

Corollary 235 *A rosette group is either D_n or C_n for some $n \geq 2$.*

Proof. Neither C_1 nor D_1 are rosette groups because neither has a non-trivial rotation. The conclusion then follows from Theorems 233 and 234. ∎

Corollary 236 *Two rosettes have the same symmetry type if and only if their respective symmetry groups are both D_n or both C_n for some $n \geq 2$.*

Exercises

1. Refer to Exercise 7.2.4 above. Which capital letters of the alphabet written in most symmetric form are rosettes?

2. For $n \geq 2$, the graph of the equation $r = \cos n\theta$ in polar coordinates is a rosette.

   (a) Find the rosette group of the graph for each $n \geq 2$.

   (b) Explain why the graph of the equation $r = \cos\theta$ in polar coordinates is *not* a rosette.

3. Find at least two rosettes in your campus architecture and determine their rosette groups.

4. If the figure is a rosette, identify its rosette group:

7.4   Frieze Patterns

Frieze patterns are typically the familiar decorative borders often seen on walls or facades, but extended infinitely far in either direction (See Figure 7.4). In this section we identify all possible symmetries of frieze patterns and arrive at the startling conclusion that frieze patterns are classified by exactly seven distinct symmetry types.

Figure 7.4. A typical frieze pattern.

Definition 237 *A translational symmetry τ of a plane figure F is a **basic translation** if it has minimal positive length, i.e., if $\tau' \in \mathrm{Sym}\,(F)$ and $\|\tau'\| > 0$, then $\|\tau\| \le \|\tau'\|$.*

Definition 238 *A **frieze pattern** is a plane figure F with the following properties:*

1. *There is a basic translation of F.*

2. *All non-trivial translational symmetries of F fix the same lines.*

*The symmetry group of a frieze pattern is called a **frieze group**.*

A frieze pattern with frieze group F_1 has exclusively translational symmetry as in Figure 7.5. The frieze group F_1 is the infinite cyclic group $\langle \tau \rangle$, where τ is a basic translation.

$$\mathbf{F\ F\ F\ F\ F\ F\ F\ F\ F\ F}$$

Figure 7.5. A frieze pattern of symmetry type F_1.

A frieze pattern with frieze group F_2 has a glide reflection symmetry γ such that γ^2 is a basic translation as in Figure 7.6. The frieze group F_2 is the infinite cyclic group $\langle \gamma \rangle$, which properly contains the infinite cyclic subgroup $\langle \gamma^2 \rangle$ of translational symmetries.

Figure 7.6. A frieze pattern of symmetry type F_2.

Although F_1 and F_2 are isomorphic as groups, they are not inner isomorphic because no inner automorphism of $\mathcal{I}$ sends $\tau \mapsto \gamma$. Thus $[F_1]$ and $[F_2]$ are distinct symmetry types.

A frieze pattern with frieze group F_3 has vertical line symmetry as in Figure 7.7. Given a line symmetry σ_l and a basic translation τ, let m be the unique line such that $\tau = \sigma_m \circ \sigma_l$. Then $\sigma_m = \tau \circ \sigma_l$ is a line symmetry since the composition of symmetries is a symmetry (Theorem 215). Furthermore, if σ_c is an arbitrary line symmetry, then $\sigma_c \circ \sigma_l = \tau^n$ for some $n \in \mathbb{Z}$, since every translational symmetry is some power of τ, and $\sigma_c = \tau^n \circ \sigma_l$. Thus the line symmetries in F_3 are the reflections $\tau^n \circ \sigma_l$, $n \in \mathbb{Z}$, and $F_3 = \langle \tau, \sigma_l \rangle$.

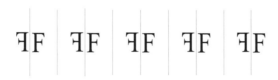

Figure 7.7. A frieze pattern of symmetry type F_3.

A frieze pattern with frieze group F_4 has point symmetry as in Figure 7.8. Given a point symmetry φ_P and a basic translation τ, let Q be the unique point such that $\tau = \varphi_Q \circ \varphi_P$. Then $\varphi_Q = \tau \circ \varphi_P$ is also a point symmetry. Furthermore, if φ_R is an arbitrary point symmetry, then $\varphi_R \circ \varphi_P = \tau^n$ for some $n \in \mathbb{Z}$, and $\varphi_R = \tau^n \circ \varphi_P$. Thus the point symmetries of F_4 are the halfturns $\tau^n \circ \varphi_P$, $n \in \mathbb{Z}$, and $F_4 = \langle \tau, \varphi_P \rangle$.

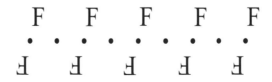

Figure 7.8. A frieze pattern of symmetry type F_4.

A frieze pattern with frieze group F_5 can be identified by its point symmetry and glide reflection symmetry as in Figure 7.9. In addition, F_5 contains vertical line symmetries, but as we shall see, such symmetries are compositions of a glide reflection with a halfturn. Let φ_P be a point symmetry and let γ be a glide reflection such that γ^2 is a basic translation. Let Q be the unique point such that $\gamma^2 = \varphi_Q \circ \varphi_P$. Then $\varphi_Q = \gamma^2 \circ \varphi_P$ is also a point symmetry. Furthermore, if φ_R is any point symmetry, $\varphi_R \circ \varphi_P = \gamma^{2n}$ for some n, and $\varphi_R = \gamma^{2n} \circ \varphi_P$.

Now the line symmetries in F_5 can be obtained from γ and φ_P as follows: Let c be the axis of γ, let b be the line through P perpendicular to c, and let a be the unique line such that $\gamma = \sigma_c \circ \sigma_b \circ \sigma_a = \varphi_P \circ \sigma_a$. Then $\sigma_a = \varphi_P \circ \gamma$ is a line symmetry. Furthermore, if σ_d is an arbitrary line symmetry, then $\sigma_d \circ \sigma_a = \gamma^{2n}$ for some $n \in \mathbb{Z}$, and $\sigma_d = \gamma^{2n} \circ \sigma_a$. Thus $F_5 = \langle \gamma, \varphi_P \rangle$.

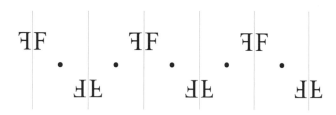

Figure 7.9. A frieze pattern of symmetry type F_5.

In Exercise 7.4.3 we ask the reader to prove that the groups F_3, F_4, and F_5 are non-abelian. Hence none of these is isomorphic to the abelian groups F_1 and F_2. Furthermore, $F_3 \not\approx_I F_4$ since no inner automorphism of $\mathcal{I}$ sends $\tau \mapsto \varphi_P$ or $\sigma_l \mapsto \varphi_P$; $F_4 \not\approx_I F_5$ since no inner automorphism of $\mathcal{I}$ sends $\tau \mapsto \gamma$ or $\varphi_P \mapsto \gamma$; and $F_3 \not\approx_I F_5$ since no inner automorphism of $\mathcal{I}$ sends $\tau \mapsto \gamma$ or $\sigma_l \mapsto \gamma$. Thus the symmetry types $[F_1], [F_2], \ldots, [F_5]$ are distinct.

A frieze pattern with symmetry group F_6 has horizontal line symmetry as in Figure 7.10. Its horizontal line symmetry σ_c is unique since a frieze pattern has translational symmetries in exactly one direction. Thus $F_6 = \langle \tau, \sigma_c \rangle$, where τ is a basic translation.

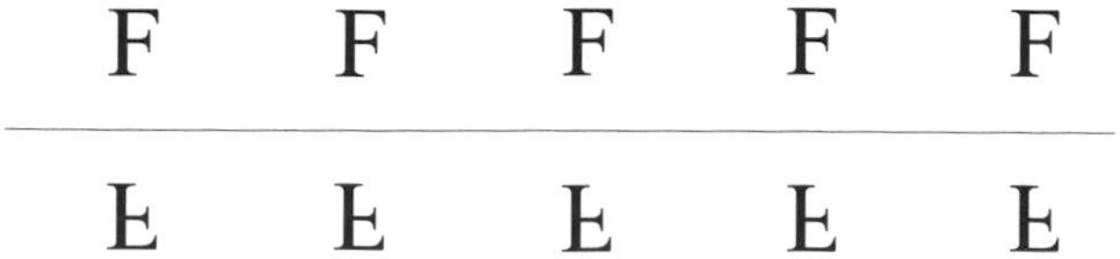

$$\mathrm{F} \qquad \mathrm{F} \qquad \mathrm{F} \qquad \mathrm{F} \qquad \mathrm{F}$$

$$\mathrm{E} \qquad \mathrm{E} \qquad \mathrm{E} \qquad \mathrm{E} \qquad \mathrm{E}$$

Figure 7.10. A frieze pattern of symmetry type F_6.

In Exercise 7.4.4 we ask the reader to prove that F_6 is abelian. Thus F_6 is not isomorphic to the non-abelian groups F_3, F_4, and F_5. Furthermore, since F_6 is not cyclic, it is not isomorphic to the cyclic groups F_1 or F_3. Thus the symmetry types $[F_1], [F_2], \ldots, [F_6]$ are distinct.

Finally, a frieze pattern with symmetry group F_7 has horizontal line symmetry and vertical line symmetry as in Figure 7.11. Let σ_c be the (unique) horizontal line symmetry, let σ_l be a vertical line symmetry, and let τ be a basic translation. If σ_m is any vertical line symmetry, then $\sigma_m \circ \sigma_l = \tau^n$ for some $n \in \mathbb{Z}$, and $\sigma_m = \tau^n \circ \sigma_l$. Let $P = c \cap l$; then $\varphi_P = \sigma_c \circ \sigma_l$ is a point symmetry. Furthermore, if φ_R is an arbitrary point symmetry, then $\varphi_R \circ \varphi_P = \tau^n$ for some $n \in \mathbb{Z}$, and $\varphi_R = \tau^n \circ \varphi_P$. Thus $F_7 = \langle \tau, \sigma_c, \sigma_l \rangle$.

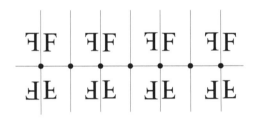

Figure 7.11. A frieze pattern of symmetry type F_7.

Since F_7 has three generators, $F_7 \not\approx F_i$ for $i = 1, 2, \ldots, 6$. Hence the symmetry types $[F_1], [F_2], \ldots, [F_7]$ are distinct.

We collect these observations in our next theorem. The proof that this list exhausts all possibilities is omitted. For a proof see [20].

Theorem 239 *There are exactly seven distinct symmetry types of frieze patterns. The seven frieze groups are:*

abelian groups	Non-abelian groups
$F_1 = \langle \tau \rangle$	$F_3 = \langle \tau, \sigma_l \rangle$
$F_2 = \langle \gamma \rangle$	$F_4 = \langle \tau, \varphi_P \rangle$
$F_6 = \langle \tau, \sigma_c \rangle$	$F_5 = \langle \gamma, \varphi_P \rangle$
	$F_7 = \langle \tau, \sigma_c, \sigma_l \rangle$

where τ is a basic translation, γ is a glide reflection symmetry such that γ^2 is a basic translation, σ_l is a vertical line symmetry, φ_P is a point symmetry, and σ_c is the unique horizontal line symmetry.

The flowchart in Figure 7.12 can be used to identify the symmetry type of a particular frieze pattern:

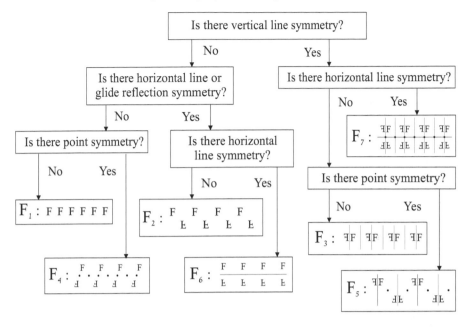

Figure 7.12. Recognition flowchart for frieze patterns.

Exercises

1. Find at least two friezes in your campus architecture and identify their frieze groups.

2. Find the frieze group of the frieze pattern in Figure 7.4.

3. Prove that the frieze groups F_3, F_4, F_5, and F_7 are non-abelian.

4. Prove that the frieze group F_6 is abelian.

5. Identify the frieze groups of the following friezes:

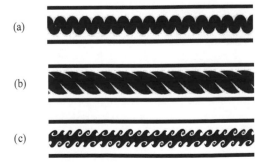

(d)

(e)

6. Identify the frieze groups of the following friezes:

(a)

(b)

(c)

(d)

(e)

(f)

(g)

(h)

(i)

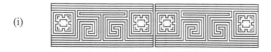

(j)

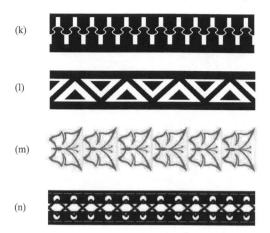

7. Identify the frieze groups of the following figures that are friezes:

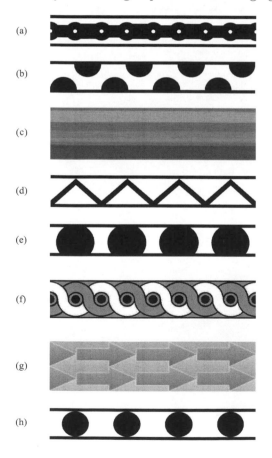

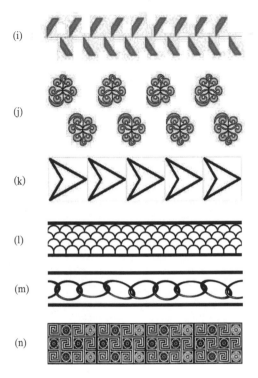

(i)

(j)

(k)

(l)

(m)

(n)

7.5 Wallpaper Patterns

A *tessellation* (or *tiling*) of the plane is a collection of congruent plane figures that fill the plane with no overlaps and no gaps. A "wallpaper pattern" is a special kind of tessellation. A typical wallpaper pattern appears in Figure 7.13.

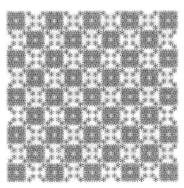

Figure 7.13. A typical wallpaper pattern.

Definition 240 *A **wallpaper pattern** is a plane figure W with translational symmetries $\tau_{\mathbf{v}}$ and $\tau_{\mathbf{w}}$, called **basic translations**, that satisfy the following properties:*

1. *The vectors $\mathbf{v}$ and $\mathbf{w}$ are linearly independent.*

2. *If τ is any translational symmetry, there exist integers m and n such that $\tau = \tau_{\mathbf{w}}^{n} \circ \tau_{\mathbf{v}}^{m} = \tau_{m\mathbf{v}+n\mathbf{w}}$.*

*The symmetry group of a wallpaper pattern is called a **wallpaper group** .*

The Penrose tiling pictured in Figure 7.14 is a tessellation with line and point symmetry but no translational symmetry. Thus the Penrose tiling is not a wallpaper pattern in the sense defined here; rather, it is a rosette with symmetry group D_5. The Penrose tiling is named after the mathematician and physicist Roger Penrose who discovered it in the 1970s.

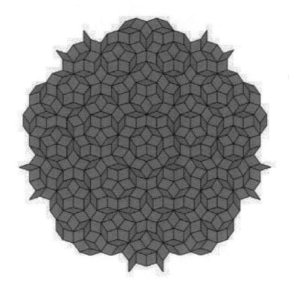

Figure 7.14. Roger Penrose's tiling has 5-fold point symmetry.

Definition 241 *Let W be a wallpaper pattern with basic translations $\tau_{\mathbf{v}}$ and $\tau_{\mathbf{w}}$. Given any point A, let $B = \tau_{\mathbf{v}}(A)$, $C = \tau_{\mathbf{w}}(B)$, and $D = \tau_{\mathbf{w}}(A)$. The **unit cell** of A is the plane region bounded by parallelogram $\square ABCD$. The **translation lattice** of A is the set of points $\mathcal{T}_A = \{(\tau_{\mathbf{w}}^{n} \circ \tau_{\mathbf{v}}^{m})(A) \mid m, n \in \mathbb{Z}\}$, which is said to be **square**, **rectangular**, or **rhombic** if the unit cell of A is **square**, **rectangular**, or **rhombic** (see Figure 7.15).*

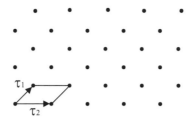

Figure 7.15. A typical translation lattice and unit cell.

Recall that a frieze pattern is a plane figure F with a non-trivial translational symmetry of shortest length. Similarly, one could define a wallpaper pattern as a plane figure with translational symmetries in independent directions and a translational symmetry of shortest length. The statement that there is a non-trivial translational symmetry of shortest length is equivalent to saying that *the length of translation is bounded away from zero*, i.e., there is some $\epsilon > 0$ such that $\|\tau\| \geq \epsilon$ for every non-trivial translation $\tau \in \mathrm{Sym}(F)$ (see Lemma 242). This latter characterization of a wallpaper pattern, which is the content of Theorem 243, is somewhat easier to apply than Definition 240 (see Exercise 7.5.4, for example).

Lemma 242 *Let F be a plane figure with non-trivial translational symmetry. Then $\mathrm{Sym}(F)$ contains a non-trivial translation of shortest length if and only if the length of translation is bounded away from zero.*

Proof. ($\Rightarrow$) If $\mathrm{Sym}(F)$ contains a non-trivial translation τ of shortest length, set $\epsilon = \|\tau\|$; then $\|\tau'\| \geq \epsilon$ for every non-trivial translation $\tau' \in \mathrm{Sym}(F)$, and the length of translation is bounded away from zero.

($\Leftarrow$) Assume that the length of translation is bounded away from zero. Then there exists an $\epsilon > 0$ such that $\|\tau\| \geq \epsilon$ for every non-trivial translation $\tau \in \mathrm{Sym}(F)$. Suppose for the sake of argument that $\mathrm{Sym}(W)$ contains no non-trivial translation of shortest length. Choose a sequence of non-trivial translational symmetries $\{\tau_k\}$ such that $\|\tau_k\| > \|\tau_{k+1}\|$ for all k. By the Monotone Convergence Theorem, there exists some $L \geq \epsilon$ and some positive integer N such that if $k > N$, then $L \leq \|\tau_k\| < L + \epsilon$. Let $\mathbf{OP}_k$ be the translation vector of τ_k; then the annulus centered at the origin with inner radius L and outer radius $L + \epsilon$ contains the points $\{P_k\}_{k>N}$. Consider the infinite collection of open disks $\{U(P_k)\}_{k>N}$, where $U(P_k) = \{X \mid XP_k < \epsilon/2\}$. Since the area of the annulus is finite, there exist distinct integers m and n such that $U(P_m)$ and $U(P_n)$ share some common point P. Then $\|\tau_n \circ \tau_m^{-1}\| = P_m P_n \leq P_m P + P P_n < \epsilon$, which is a contradiction. Therefore $\mathrm{Sym}(W)$ contains a non-trivial translation of shortest length. ∎

Theorem 243 *Let W be a plane figure with non-trivial translational symmetries in independent directions. Then W is a wallpaper pattern if and only if the length of translation is bounded away from zero.*

Proof. ($\Rightarrow$) If W is a wallpaper pattern, choose basic translations $\tau_{\mathbf{v}}$ and $\tau_{\mathbf{w}}$, and consider the translation lattice $\mathcal{T}_O$. Let $U(O)$ denote the open disk centered at the origin O of radius $1 + \|\mathbf{v}\|$. Since $U(O)$ has finite area, it contains finitely many points of $\mathcal{T}_O$. Let $\epsilon = \min\{OP \mid P \in U(O) \cap \mathcal{T}_O \smallsetminus \{O\}\}$. If $\tau \in \text{Sym}(W)$ is a non-trivial translation, then $\tau(O) \in \mathcal{T}_O \smallsetminus \{O\}$ and $\|\tau\| \geq \epsilon$. Hence the length of translation is bounded away from zero.

($\Leftarrow$) Assume that the length of translation is bounded away from zero. By Lemma 242, there is a non-trivial translation $\tau_{\mathbf{v}} \in \text{Sym}(W)$ of shortest length. Since $\text{Sym}(W)$ contains non-trivial translations in independent directions, consider the set $T \subset \text{Sym}(W)$ of non-trivial translations in directions not parallel to $\mathbf{v}$. By Lemma 242, there is a non-trivial translation $\tau_{\mathbf{w}} \in T$ of shortest length. We claim that $\tau_{\mathbf{v}}$ and $\tau_{\mathbf{w}}$ are basic translations.

Let $\tau \in \text{Sym}(W)$ and let $X = \tau(O)$. Then X lies either on or in the interior of some parallelogram $\square ABCD$, where $A = \left(\tau_{\mathbf{w}}^{j} \circ \tau_{\mathbf{v}}^{i}\right)(O)$ for some $i, j \in \mathbb{Z}$, $B = \tau_{\mathbf{v}}(A)$, $C = \tau_{\mathbf{w}}(B)$, and $D = \tau_{\mathbf{w}}(A)$. Note that $\tau_{\mathbf{AX}}, \tau_{\mathbf{BX}}, \tau_{\mathbf{DX}} \in \text{Sym}(W)$, since $\mathbf{AX} = \mathbf{OX} - \mathbf{OA}$, $\mathbf{BX} = \mathbf{AX} - \mathbf{AB}$, and $\mathbf{DX} = \mathbf{AX} - \mathbf{AD}$. Furthermore, each of these translations is non-trivial whenever X is not a vertex of $\square ABCD$.

Suppose X is not a vertex of $\square ABCD$. Then X is off $\overline{AB}$ and $\overline{CD}$ since $\tau_{\mathbf{v}}$ is a non-trivial translation of shortest length, and X is off $\overline{AD}$ and $\overline{BC}$ since $\tau_{\mathbf{w}}$ is a non-trivial translation of shortest length in directions not parallel to $\mathbf{v}$. Thus X is an interior point of $\square ABCD$. Consequently, neither $\mathbf{BX}$ nor $\mathbf{DX}$ is parallel to $\mathbf{v}$, and $\|\mathbf{BX}\| + \|\mathbf{DX}\| < \|\mathbf{v}\| + \|\mathbf{w}\|$, as the reader can easily check. But $\min\{\|\mathbf{BX}\|, \|\mathbf{DX}\|\} \leq \frac{1}{2}(\|\mathbf{BX}\| + \|\mathbf{DX}\|) < \frac{1}{2}(\|\mathbf{v}\| + \|\mathbf{w}\|) \leq \|\mathbf{w}\|$ contradicts the fact that $\tau_{\mathbf{w}}$ is a non-trivial translation of shortest length in directions not parallel to $\mathbf{v}$. Therefore X is a vertex of $\square ABCD$, and it follows that $\tau = \tau_{\mathbf{w}}^{n} \circ \tau_{\mathbf{v}}^{m}$ for some $m, n \in \mathbb{Z}$. Thus $\tau_{\mathbf{v}}$ and $\tau_{\mathbf{w}}$ are basic translations and W is a wallpaper pattern by definition. $\blacksquare$

In retrospect, we see that parallelogram $\square ABCD$ in the proof of Theorem 243 is the unit cell of A in the wallpaper pattern W.

Definition 244 *Let $n \geq 2$. A point P is an n-**center** of a wallpaper pattern if the subgroup of rotational symmetries centered at P is C_n.*

Example 245 Each vertex of a hexagonal tessellation, or "honeycomb," is a 3-center and the centroid of each hexagon is a 6-center (see Figure 7.16).

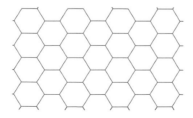

Figure 7.16. A honeycomb tessellation has 3-centers and 6-centers.

Theorem 246 *The symmetries of a wallpaper pattern W preserve n-centers, i.e., if $\alpha \in \mathrm{Sym}\,(W)$ and P is an n-center of W, then so is $\alpha\,(P)$.*

Proof. Let P be an n-center of W, let $\alpha \in \mathrm{Sym}\,(W)$, and let $Q = \alpha\,(P)$. Consider the subgroup of rotational symmetries $\langle \rho_{P,\Theta} \rangle$, where $\Theta = \frac{360}{n}$. Then by conjugation and closure we have $\alpha \circ \rho_{P,\Theta} \circ \alpha^{-1} = \rho_{Q,\pm\Theta} \in \mathrm{Sym}\,(W)$ and consequently,

$$\rho_{Q,\pm\Theta}^{n} = \left(\alpha \circ \rho_{P,\Theta} \circ \alpha^{-1}\right)^{n} = \alpha \circ \rho_{P,\Theta}^{n} \circ \alpha^{-1} = \iota.$$

Hence $\rho_{Q,\Theta}^{n} = \iota$ or $\rho_{Q,-\Theta}^{n} = \iota$. But if $\rho_{Q,-\Theta}^{n} = \iota$, we also have $\rho_{Q,\Theta}^{n} = \rho_{Q,-(-\Theta)}^{n} = \rho_{Q,-\Theta}^{-n} = \left(\rho_{Q,-\Theta}^{n}\right)^{-1} = \iota^{-1} = \iota$. So $\rho_{Q,\Theta}^{n} = \iota$ in either case. Let m be the smallest positive integer such that $\rho_{Q,\Theta}^{m} = \iota$. Then Q is an m-center and $m \leq n$. Now mirror this argument starting with the m-center Q, the symmetry α^{-1}, and $P = \alpha^{-1}\,(Q)$ to conclude that $\rho_{P,\Theta}^{m} = \iota$. Since P is an n-center, $n \leq m$. Therefore $m = n$ and Q is an n-center. ∎

Two n-centers in a wallpaper pattern cannot be arbitrarily close to each other.

Theorem 247 *Let A and B be distinct n-centers in a wallpaper pattern W and let $\tau \in \mathrm{Sym}\,(W)$ be a translation of shortest length. Then $AB \geq \frac{1}{2}\,\|\tau\|$.*

Proof. Let $\tau_{\mathbf{v}}$ and $\tau_{\mathbf{w}}$ be basic translations and let $\Theta = \frac{360}{n}$. Then $\rho_{B,\Theta}\circ\rho_{A,-\Theta}$ is a non-trivial translational symmetry by closure and the Angle Addition Theorem. Hence there exist integers i and j, not both zero, such that $\rho_{B,\Theta} \circ \rho_{A,-\Theta} = \tau_{\mathbf{w}}^{j} \circ \tau_{\mathbf{v}}^{i}$ and $\rho_{B,\Theta} = \tau_{\mathbf{w}}^{j} \circ \tau_{\mathbf{v}}^{i} \circ \rho_{A,\Theta}$. Consider the point

$$A_{ij} = \left(\tau_{\mathbf{w}}^{j} \circ \tau_{\mathbf{v}}^{i}\right)(A) = \left(\tau_{\mathbf{w}}^{j} \circ \tau_{\mathbf{v}}^{i} \circ \rho_{A,\Theta}\right)(A) = \rho_{B,\Theta}\,(A) \neq A.$$

Then $AA_{ij} \geq \|\tau\|$, since the translation $\tau_{\mathbf{w}}^{j} \circ \tau_{\mathbf{v}}^{i} \neq \iota$ and τ is a translation of shortest length, and $AB = BA_{ij}$ since A and A_{ij} are equidistant from the center of rotation B. Furthermore, $AB + BA_{ij} \geq AA_{ij}$ by the triangle inequality. Therefore $2AB = AB + BA_{ij} \geq AA_{ij} \geq \|\tau\|$. ∎

Our next theorem, which was first proved by the Englishman W. Barlow in the late 1800s, is quite surprising. It tells us that wallpaper patterns cannot have 5-centers; consequently, *crystalline structures cannot have pentagonal symmetry.*

Theorem 248 (Crystallographic Restriction) *If P is an n-center of a wallpaper pattern, then $n \in \{2, 3, 4, 6\}$.*

Proof. Let P be an n-center and let τ be a translation of shortest length.

Claim. *There is an n-center Q distinct from and closest to P.* Suppose, on the contrary, that no such Q exists and consider any n-center $Q_1 \neq P$. Since PQ_1 is not an n-center closest to P, there is an n-center $Q_2 \neq P$ such that

$PQ_1 > PQ_2$. Inductively, there is an infinite sequence of n-centers $\{Q_k\}$ distinct from P such that $PQ_1 > PQ_2 > \cdots$. But $PQ_k \geq \frac{1}{2} \|\tau\|$ for all k by Theorem 247. Hence $\{PQ_k\}$ is a strictly decreasing sequence of positive real numbers, which converges to $M \geq \frac{1}{2} \|\tau\|$, i.e., given $\epsilon > 0$, there is a positive integer N such that if $k > N$ then $M < PQ_k < M + \epsilon$. Consequently, infinitely many n-centers Q_k lie in the annulus S centered at P with inner radius M and outer radius $M + \epsilon$. But this is impossible since S has finite area and $Q_i Q_j \geq \frac{1}{2} \|\tau\|$ for all i, j, which proves the claim.

Choose an n-center Q distinct from and closest to P and let $\Theta = \frac{360}{n}$. Then $R = \rho_{Q,\Theta}(P)$ and $S = \rho_{R,\Theta}(Q)$ are n-centers by Theorem 246, and $PQ = QR = RS$. If $S = P$, then $\triangle PQR$ is equilateral and $P = \rho_{R,\Theta}(Q)$, in which case $\Theta = 60$ and $n = 6$. If $S \neq P$, then $SP \geq PQ$ by the choice of Q, in which case $\Theta \geq 90$ and $n \leq 4$. Therefore $n \in \{2, 3, 4, 6\}$. ∎

Corollary 249 *A wallpaper pattern with a 4-center has no 3 or 6-centers.*

Proof. Let W be a wallpaper pattern. If P is a 3-center and Q is a 4-center, then $\rho_{P,120}, \rho_{Q,-90} \in \mathrm{Sym}(W)$ and $\rho_{P,120} \circ \rho_{Q,-90} = \rho_{S,30} \in \mathrm{Sym}(W)$ by closure. Thus W has an n-center S with $n \geq 12$, contradicting Theorem 248. Similarly, if Q is a 4-center and R is a 6-center, then $\rho_{Q,90}, \rho_{R,-60} \in \mathrm{Sym}(W)$ and $\rho_{R,-60} \circ \rho_{Q,90} = \rho_{S,30} \in \mathrm{Sym}(W)$ by closure. Again, W has an n-center S with $n \geq 12$, which is a contradiction. ∎

In addition to translational symmetry, wallpaper patterns can have line symmetry, glide reflection symmetry, and point symmetries of 180°, 120°, 90°, or 60°. Since the only point symmetries in a frieze group are halfturns, it is not surprising to find more wallpaper groups than frieze groups. In fact, there are exactly seventeen!

We use the international standard notation to denote the various wallpaper groups. Each symbol is a string of letters and integers selected from p, c, m, g and $1, 2, 3, 4, 6$. The letter p stands for *primitive translation lattice*. The points in a primitive translation lattice are the vertices of parallelograms with no interior points of symmetry. When *a point of symmetry lies at the centroid of some unit cell*, we use the letter c. The letter m stands for *mirror* and indicates a *line of symmetry*; the letter g stands for *glide* and indicates a *glide reflection symmetry*. Integers indicate the *maximal order* of the rotational symmetries in a wallpaper group. Let W denote a wallpaper pattern.

There are four symmetry types of wallpaper patterns with *no n-centers*.

- If W has *no lines of symmetry or glide reflection symmetry*, $\mathrm{Sym}(W) = p1$.

- If W has *glide reflection symmetry but no lines of symmetry*, $\mathrm{Sym}(W) = pg$.

- If W has both *lines of symmetry and glide reflection symmetry*, and

  - *the axis of some glide reflection symmetry is not a line of symmetry,* $\operatorname{Sym}(W) = cm$.
  - *the axis of every glide reflection symmetry is a line of symmetry,* $\operatorname{Sym}(W) = pm$.

There are five symmetry types of wallpaper patterns whose *n-centers are all 2-centers*.

- If W has *no lines of symmetry or glide reflection symmetry*, $\operatorname{Sym}(W) = p2$.

- If W has *glide reflection symmetry but no lines of symmetry*, $\operatorname{Sym}(W) = pgg$.

- If W has *lines of symmetry in one direction*, $\operatorname{Sym}(W) = pmg$.

- If W has *lines of symmetry in two directions*, and

  - *each 2-center is on a line of symmetry*, $\operatorname{Sym}(W) = pmm$.
  - *some 2-center is off the lines of symmetry*, $\operatorname{Sym}(W) = cmm$.

Three symmetry types of wallpaper patterns have *n-centers whose smallest rotation angle is* $90°$.

- If W has *no lines of symmetry*, $\operatorname{Sym}(W) = p4$.

- If W has *lines of symmetry in four directions*, $\operatorname{Sym}(W) = p4m$.

- If W has *lines of symmetry, but not in four directions*, $\operatorname{Sym}(W) = p4g$.

Three symmetry types of wallpaper patterns have *n-centers whose smallest rotation angle is* $120°$.

- If W has *no lines of symmetry*, $\operatorname{Sym}(W) = p3$.

- If W has *lines of symmetry*, and

  - *each 3-center is on some line of symmetry*, $\operatorname{Sym}(W) = p3m1$.
  - *some 3-center is off the lines of symmetry*, $\operatorname{Sym}(W) = p31m$.

Finally, two symmetry types of wallpaper patterns have *n-centers whose smallest rotation angle is* $60°$.

- If W has *no lines of symmetry*, $\operatorname{Sym}(W) = p6$.

- If W has *lines of symmetry*, $\operatorname{Sym}(W) = p6m$.

We collect the remarks above in our next theorem. The proof that this list exhausts all possibilities is omitted. For a proof see [20].

Theorem 250 *There are exactly seventeen distinct symmetry types of wallpaper patterns. The seventeen wallpaper groups are:*

p1	p2	p3	p4	p6
pg	pgg	p31m	p4g	p6m
pm	pmg	p3m1	p4m	
cm	cmm			
	pmm			

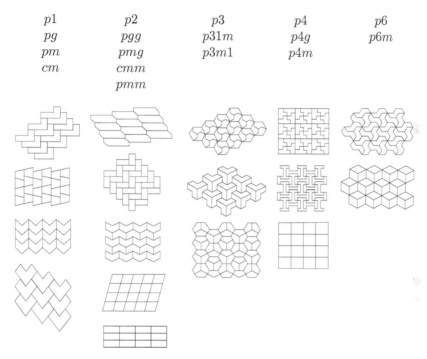

The recognition flowchart for wallpaper patterns in Figure 7.17 can be used to identify the wallpaper group of a given wallpaper pattern.

Example 251 Try your hand at identifying the wallpaper groups of the following wallpatterns:

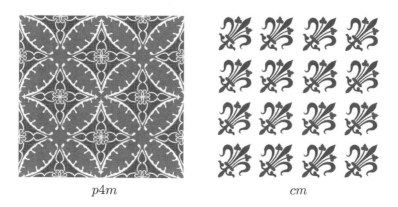

p4m   cm

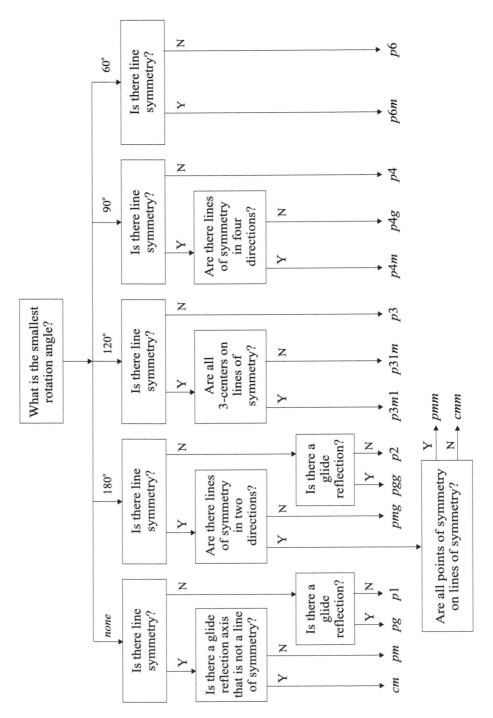

Figure 7.17. Recognition flowchart for wallpaper patterns.

We conclude our discussion of wallpaper patterns with a brief look at "edge tessellations," which have a simple and quite beautiful classification.

Definition 252 *An **edge tessellation** is a tessellation of the plane generated by reflecting a polygon in the lines containing its edges.*

Obviously, regular hexagons, rectangles, and equilateral, 60-right and isosceles right triangles generate edge tessellations, but are there others? The complete list was discovered by Millersville University students Andrew Hall, Joshua York, and Matthew Kirby in the spring of 2009. We present their result here as Theorem 253, and leave the proof, which follows easily from the Crystallographic Restriction, as an exercise for the reader (see Exercise 7.5.5).

If a polygon P generates an edge tessellation, the edges of P are on lines of symmetry, and consequently, the vertices of P are points of symmetry. Since P has at least three non-collinear vertices, and a plane figure with two distinct points of symmetry has non-trivial translational symmetry, there are translational symmetries in at least two directions (see Exercise 7.5.4). Furthermore, the length of translation in a given direction is at least at long as the longest cross-section of P in that direction. Consequently, the length of translation is bounded away from zero and edge tessellations are wallpaper patterns.

Let V be a vertex of P. What are the possible values of $m\angle V$ (the interior angle at V)? The bisector m of $\angle V$ either is or is not a line of symmetry in the edge tessellation (if P is an equilateral triangle, for example, then m is a line of symmetry; but if P is a rectangle that is not a square, then m is not a line of symmetry). So suppose m is a line of symmetry. Reflecting P in an edge containing V, then reflecting the image of P in m rotates P about V through the angle $m\angle V$. Since V is an n-center, and $n = 2, 3, 4, 6$ by the Crystallographic Restriction, $m\angle V \in \{180, 120, 90, 60\}$. Since the interior angles of P are never straight angles, $m\angle V \in \{120, 90, 60\}$. On the other hand, suppose m is not a line of symmetry. Let e and e' be the edges of P sharing vertex V. Reflecting P in e, then reflecting the image of P in e' rotates P about V through the angle $2m\angle V$. Thus $2m\angle V = \{180, 120, 90, 60\}$ and $m\angle V \in \{90, 60, 45, 30\}$. Combining the possibilities in each case, we conclude that $m\angle V \in \{120, 90, 60, 45, 30\}$.

Theorem 253 *A polygon generating an edge tessellation is one of the following eight types: a rectangle; an equilateral, 60-right, isosceles right, or 120-isosceles triangle; a 120-rhombus; a 60-90-120 kite; or a regular hexagon.*

Thus four edge tessellations are generated by non-obtuse polygons and four are generated by non-obtuse polygons (see Figures 7.18 and 7.19). Exactly three of the 17 symmetry types of wallpaper patterns appear as edge tessellations. Non-square rectangles generate edge tessellations with wallpaper group *pmm*, isosceles right triangles and squares generate edge tessellations with wallpaper group *p4m*, and the remaining six polygons in Theorem 253 generate edge tessellations with wallpaper group *p6m*.

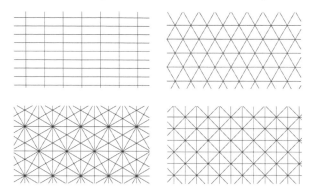

Figure 7.18. Edge tessellations generated by non-obtuse polygons.

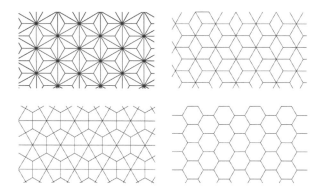

Figure 7.19. Edge tessellations generated by obtuse polygons.

In [3] Andrew Baxter and the first author applied the edge tessellation generated by an equilateral triangle to find and classify the periodic orbits of billiards on an equilateral triangle. In unpublished work, Baxter and undergraduates Ethan McCarthy and Jonathan Eskreis-Winkler used the same technique to find and classify the periodic orbits on a square, a rectangle, a 60-right, and an isosceles right triangle. At the time of this writing, the four obtuse cases remain open.

Exercises

1. Identify the wallpaper group of the wallpaper pattern pictured in Figure 7.13.

2. Find at least two different wallpaper patterns on your campus and identify their wallpaper groups.

3. Prove that if a plane figure F has distinct points of symmetry A and B, then Sym (F) contains a non-trivial translation and, consequently, has infinite order.

4. Let P be a polygon that generates an edge tessellation $\mathcal{T}$. Prove that $\mathcal{T}$ has translational symmetries in at least two directions.

5. Prove Theorem 253: A polygon generating an edge tessellation is one of the following eight types: a rectangle; an equilateral, 60-right, isosceles right, or 120-isosceles triangle; a 120-rhombus; a 60-90-120 kite; or a regular hexagon.

6. Identify the wallpaper groups of the following wallpaper patterns:

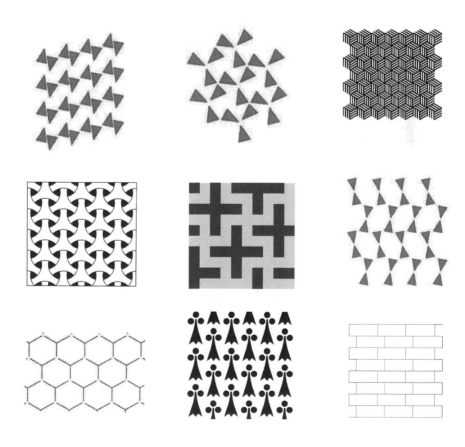

7. Identify the wallpaper groups of the following patterns:

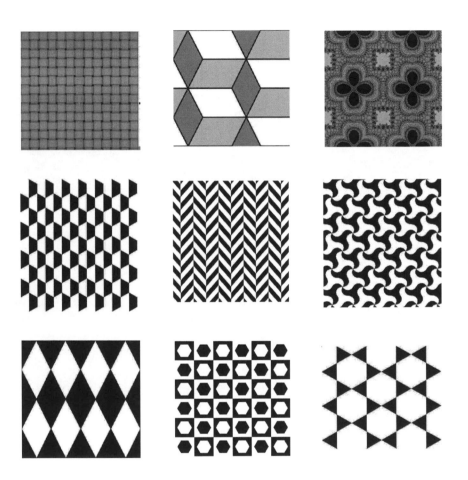

Chapter 8

Similarity

Understand that a two-dimensional figure is similar to another if the second can be obtained from the first by a sequence of rotations, reflections, translations, and dilations; given two similar two-dimensional figures, describe a sequence that exhibits the similarity between them.

Common Core State Standards for Mathematics
CCSS.MATH.CONTENT.8.G.A.4

In this chapter we consider transformations that magnify or stretch the plane. Such transformations are called "similarities" or "size transformations." One uses a similarity to relate two similar triangles in much the same way one uses an isometry to relate two congruent triangles. The "stretch" similarities, which linearly expand the plane radially outward from some fixed point, are especially important because they provide an essential component of every non-isometric similarity. Indeed, we shall prove that every similarity is one of the following four distinct similarity types: an isometry, a stretch, a stretch rotation, or a stretch reflection.

8.1 Plane Similarities

Recall Definition 93: A **similarity of ratio** $r > 0$ is a transformation $\alpha : \mathbb{R}^2 \to \mathbb{R}^2$ such that if P and Q are points, $P' = \alpha(P)$, and $Q' = \alpha(Q)$, then $P'Q' = rPQ$.

Note that if $r = 1$, then $P'Q' = PQ$ and α is an isometry. Furthermore, if α has distinct fixed points P and Q, then $P' = P$, $Q' = Q$, and $P'Q' = PQ$, the ratio of similarity $r = 1$ and α is an isometry. Thus α is the identity or a reflection by Theorem 170. Of course, if α has three non-collinear fixed points, then $\alpha = \iota$ by Theorem 135. This proves:

Proposition 254 *A similarity of ratio 1 is an isometry; a similarity with two or more distinct fixed points is a reflection or the identity; a similarity with three non-collinear fixed points is the identity.*

Similarities are bijective by Proposition 95. In fact:

Proposition 255 *The set S of all similarities is a group with respect to compositions.*

Proof. The proof is left to the reader. ∎

Corollary 256 (Three Points Theorem for Similarities) *Two similarities that agree on three non-collinear points are equal.*

Proof. Let α and β be similarities, and let A, B, and C be non-collinear points such that

$$\alpha(A) = \beta(A), \ \alpha(B) = \beta(B), \text{ and } \alpha(C) = \beta(C). \tag{8.1}$$

Compose β^{-1} with both sides of each equation in line (8.1) and obtain

$$\left(\beta^{-1} \circ \alpha\right)(A) = A, \ \left(\beta^{-1} \circ \alpha\right)(B) = B, \text{ and } \left(\beta^{-1} \circ \alpha\right)(C) = C.$$

But by Proposition 255, $\beta^{-1} \circ \alpha$ is a similarity with three non-collinear fixed points. Hence $\beta^{-1} \circ \alpha = \iota$ by Proposition 254, and $\alpha = \beta$. ∎

Definition 257 *Let C and P be points, and let $r > 0$. A **stretch about** C **of ratio** r is the transformation $\xi_{C,r} : \mathbb{R}^2 \to \mathbb{R}^2$ with the following properties:*

1. $\xi_{C,r}(C) = C$.

2. *If $P \neq C$, then $P' = \xi_{C,r}(P)$ is the unique point on $\overrightarrow{CP}$ such that $CP' = rCP$.*

*The point C is called the **center** of the stretch (see Figure 8.1).*

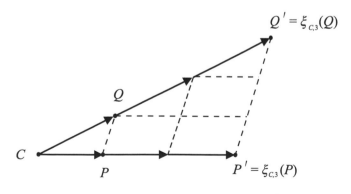

Figure 8.1. A stretch about C of ratio 3.

Note that if $D = \xi_{C,r}(C)$, then $CD = rCC = 0$ and $D = C$. Thus a stretch fixes its center. Of course, the identity is a stretch of ratio 1 about every point C. Furthermore, the equations of a stretch about the origin are

$$\xi_{O,r} : \begin{cases} x' = rx \\ y' = ry. \end{cases}$$

To obtain the equations of a stretch about $C = \begin{bmatrix} a \\ b \end{bmatrix}$ of ratio r, conjugate $\xi_{O,r}$ by the translation τ_{OC}, i.e.,

$$\xi_{C,r} = \tau_{OC} \circ \xi_{O,r} \circ \tau_{OC}^{-1}.$$

Composing equations we obtain:

Proposition 258 *Let $C = \begin{bmatrix} a \\ b \end{bmatrix}$ and let $r > 0$. The equations of $\xi_{C,r}$ are*

$$\begin{cases} x' = rx + (1 - r)\, a \\ y' = ry + (1 - r)\, b. \end{cases}$$

A stretch acts on vectors in the following way:

Proposition 259 *Let $\mathbf{v}$ be a vector, and let $r > 0$. Then $\xi_{C,r}(\mathbf{v}) = r\mathbf{v}$.*

Proof. Let $C = \begin{bmatrix} a \\ b \end{bmatrix}$ and let $\mathbf{v} = \mathbf{PQ}$ where $P = \begin{bmatrix} p_1 \\ p_2 \end{bmatrix}$ and $Q = \begin{bmatrix} q_1 \\ q_2 \end{bmatrix}$. Then

$$P' = \xi_{C,r}(P) = \begin{bmatrix} rp_1 + (1-r)\,a \\ rp_2 + (1-r)\,b \end{bmatrix} \text{ and } Q' = \xi_{C,r}(Q) = \begin{bmatrix} rq_1 + (1-r)\,a \\ rq_2 + (1-r)\,b \end{bmatrix}$$

so that

$$\xi_{C,r}(\mathbf{v}) = \mathbf{P'Q'} = \begin{bmatrix} r(q_1 - p_1) \\ r(q_2 - p_2) \end{bmatrix} = r\mathbf{v}.$$

∎

Corollary 260 *Let C be a point and let $r > 0$. Then the stretch $\xi_{C,r}$ is a similarity of ratio r.*

Proof. Let C be a point and let $r > 0$. Let P and Q be two points, $P' = \xi_{C,r}(P)$, and $Q' = \xi_{C,r}(Q)$. Then $\mathbf{P'Q'} = \xi_{C,r}(\mathbf{PQ})$ so that $\mathbf{P'Q'} = r\mathbf{PQ}$ by Proposition 259. Therefore $P'Q' = rPQ$ and $\xi_{C,r}$ is a similarity of ratio r. ∎

Exercises

1. One can use the following procedure to determine the height of an object: Place a mirror flat on the ground and move back until you can see the top of the object in the mirror. Explain how this works.

2. Find the ratio of similarity r for a similarity α such that $\alpha\left(\begin{bmatrix}1\\2\end{bmatrix}\right) = \begin{bmatrix}0\\0\end{bmatrix}$ and $\alpha\left(\begin{bmatrix}3\\4\end{bmatrix}\right) = \begin{bmatrix}3\\4\end{bmatrix}$.

3. Find the point P and ratio of similarity r such that $\xi_{P,r}\left(\begin{bmatrix}x\\y\end{bmatrix}\right) = \begin{bmatrix}3x+7\\3y-5\end{bmatrix}$.

4. Let $C = \begin{bmatrix}a\\b\end{bmatrix}$ be a point, let $\Theta \in \mathbb{R}$, and let $r > 0$. Use the equations of $\rho_{C,\Theta}$ and $\xi_{C,r}$ to prove that $\rho_{C,\Theta} \circ \xi_{C,r} = \xi_{C,r} \circ \rho_{C,\Theta}$.

5. Let $C = \begin{bmatrix}a\\b\end{bmatrix}$ be a point on line $m : cX + dY + e = 0$, $c^2 + d^2 > 0$, and let $r > 0$. Use the equations of σ_m and $\xi_{C,r}$ to prove that $\sigma_m \circ \xi_{C,r} = \xi_{C,r} \circ \sigma_m$.

6. Prove Proposition 255: The set $\mathcal{S}$ of all similarities is a group with respect to composition.

7. If α is a similarity of ratio r and A, B, and C are distinct non-collinear points, let $A' = \alpha(A)$, $B' = \alpha(B)$, and $C' = \alpha(C)$. Prove that $\angle ABC \cong \angle A'B'C'$.

8. Prove that similarities preserve betweenness.

8.2   Classification of Dilatations

In this section we take a closer look at the family of dilatations introduced in Section 4.1, and observe that every dilatation is either a translation, a stretch, or a stretch followed by a halfturn. We begin with the following proposition, which asserts that a stretch is a dilatation.

Proposition 261 *Let C be a point and let $r > 0$. Then the stretch $\xi_{C,r}$ is a dilatation.*

Proof. Let m be a line. If C is on m, then $\xi_{C,r}(m) = m$ by definition, and $\xi_{C,r}(m) \parallel m$. So assume C is off m, and choose distinct points P, Q, and R on m (see Figure 8.2). By definition, $CP' = rCP$, $CQ' = rCQ$, and $CR' = rCR$. Thus $\frac{CP}{CP'} = \frac{CQ}{CQ'}$ so that $\overrightarrow{PQ} \parallel \overrightarrow{P'Q'}$ by Theorem 64, and similarly, $\overrightarrow{PR} \parallel \overrightarrow{P'R'}$. Since $\overleftrightarrow{P'Q'}$ and $\overleftrightarrow{P'R'}$ are both parallel to line $m = \overleftrightarrow{PQ} = \overleftrightarrow{PR}$. By the Euclidean Parallel Postulate, $\overleftrightarrow{P'Q'} = \overleftrightarrow{P'R'}$. Consequently, P', Q', and R' are collinear, and $\xi_{C,r}(m) = \overleftrightarrow{P'Q'} = \overleftrightarrow{P'R'}$. Therefore $\xi_{C,r}$ is a collineation. But $\xi_{C,r}$ is also a dilatation since $m \parallel \xi_{C,r}(m)$. $\blacksquare$

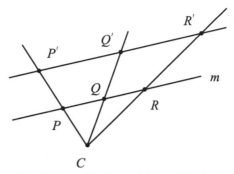

Figure 8.2. Points P, Q, and R and their images.

Definition 262 *Let C be a point and let $r > 0$. A **dilation about C of ratio r**, denoted by $\delta_{C,r}$, is either a stretch about C of ratio r or a stretch about C of ratio r followed by a halfturn about C. The point C is called the **center of dilation**.*

Note that the identity and all halfturns are dilations of ratio 1. If $C-P-Q$, there is a positive real number r such that $Q = \delta_{C,r}(P) = \xi_{C,r}(P)$ and if $P - C - Q$, there is a positive real number r such that $Q = \delta_{C,r}(P) = (\varphi_C \circ \xi_{C,r})(P)$.

Theorem 263 *Let C be a point and let $r > 0$. Then the dilation $\delta_{C,r}$ is both a dilatation and a similarity of ratio r.*

Proof. If $\delta_{C,r} = \xi_{C,r}$, the conclusion follows from Corollary 260 and Proposition 261. So assume that $\delta_{C,r} = \varphi_C \circ \xi_{C,r}$. A halfturn is a similarity since it is an isometry; it is also a dilatation by Proposition 125. So on one hand, $\varphi_C \circ \xi_{C,r}$ is a composition of dilatations, which is a dilatation by Exercise 7.1.5, and on the other hand $\varphi_C \circ \xi_{C,r}$ is a composition of similarities, which is a similarity by Proposition 255. Proof of the fact that $\varphi_C \circ \xi_{C,r}$ is a similarity of ratio r is left to the reader. ∎

Theorem 264 *If $\overleftrightarrow{AB} \parallel \overleftrightarrow{DE}$, there is a unique dilatation α such that $D = \alpha(A)$ and $E = \alpha(B)$.*

Proof. If $A = D$ and $B = E$, set $\alpha = \iota$. Otherwise, we define a dilatation with the required property, then verify uniqueness. Let $C = \tau_{AD}(B)$, let $r = DE/DC$, and consider the dilation $\delta_{D,r}$ such that $E = \delta_{D,r}(C)$. Then

$$(\delta_{D,r} \circ \tau_{AD})(A) = \delta_{D,r}(D) = D \text{ and } (\delta_{D,r} \circ \tau_{AD})(B) = \delta_{D,r}(C) = E.$$

Since τ_{AD} and $\delta_{D,r}$ are dilatations, $\delta_{D,r} \circ \tau_{AD}$ is a dilatation with the required property.

For uniqueness, let α be any dilatation such that $\alpha(A) = D$ and $\alpha(B) = E$. Let P be a point off $\overleftrightarrow{AB}$. To locate $P' = \alpha(P)$, note that P' is on the line l

through D parallel to $\overleftrightarrow{AP}$ as well as the line m through E parallel to $\overleftrightarrow{BP}$ (see Figure 8.3). Hence $P' = l \cap m$. Let Q be a point on $\overleftrightarrow{AB}$ distinct from A and let $Q' = \alpha(Q)$. Then Q' is on the line $\overleftrightarrow{DE}$ as well as the line n through P' parallel to $\overleftrightarrow{PQ}$. Hence $Q' = \overleftrightarrow{DE} \cap n$. In either case, P' and Q' are uniquely determined by the given points A, B, D, and E. Thus every dilatation with the required property sends P to P' and Q to Q', and in particular,

$$\alpha(A) = D = (\delta_{D,r} \circ \tau_{\mathbf{AD}})(A),$$
$$\alpha(P) = P' = (\delta_{D,r} \circ \tau_{\mathbf{AD}})(P),$$
$$\alpha(Q) = Q' = (\delta_{D,r} \circ \tau_{\mathbf{AD}})(Q).$$

Therefore $\alpha = \delta_{D,r} \circ \tau_{\mathbf{AD}}$ by Corollary 256. ∎

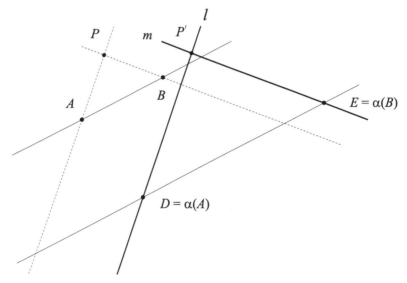

Figure 8.3. Lines l through D parallel to $\overleftrightarrow{AP}$ and m through E parallel to $\overleftrightarrow{BP}$.

Proposition 265 *If α is a dilatation, $B = \alpha(A)$, and $B \neq A$, then α fixes $\overleftrightarrow{AB}$.*

Proof. Let $C = \alpha(B)$. Since α is a dilatation, $\overleftrightarrow{AB} \parallel \overleftrightarrow{BC}$. Therefore $\overleftrightarrow{AB} = \overleftrightarrow{BC} = \alpha\left(\overleftrightarrow{AB}\right)$. ∎

We can now determine all dilatations.

Theorem 266 (Classification of Dilatations) *A dilatation is a translation, a halfturn or a dilation.*

Proof. Let α be a dilatation. If α is an isometry, it is either a translation or a halfturn by Theorem 108, Proposition 125, and Exercise 6.2.8. So assume α is a non-isometric dilatation. Let A, B, and C be non-collinear points, and let $A' = \alpha(A)$, $B' = \alpha(B)$, and $C' = \alpha(C)$. Since dilatations are collineations and collineations are bijective, A', B', and C' are distinct and $\overleftrightarrow{AB} \parallel \overleftrightarrow{A'B'}$, $\overleftrightarrow{AC} \parallel \overleftrightarrow{A'C'}$, and $\overleftrightarrow{BC} \parallel \overleftrightarrow{B'C'}$. Consequently, corresponding interior angles of $\triangle ABC$ and $\triangle A'B'C'$ are congruent and $\triangle ABC \sim \triangle A'B'C'$. But $\triangle ABC \ncong \triangle A'B'C'$ since α is not an isometry, and the lines in at least one of $\overleftrightarrow{AB} \parallel \overleftrightarrow{A'B'}$, $\overleftrightarrow{AC} \parallel \overleftrightarrow{A'C'}$, and $\overleftrightarrow{BC} \parallel \overleftrightarrow{B'C'}$ are distinct. Without loss of generality, assume that $\overleftrightarrow{AB} \neq \overleftrightarrow{A'B'}$. Then by Exercise 8.2.2, $D = \overleftrightarrow{AA'} \cap \overleftrightarrow{BB'} = \alpha(D)$, which is off $\overleftrightarrow{AB}$ and $\overleftrightarrow{A'B'}$ since A, A', and D are distinct and collinear (see Figures 8.4 and 8.5). Thus $\triangle ABD \sim \triangle A'B'D$ (by AA) with ratio of similarity $r = DA'/DA = DB'/DB \neq 1$. Thus $DA' = rDA$ and $DB' = rDB$, and either $D - A - A'$, $D - A' - A$, or $A - D - A'$. If $D - A - A'$, then $r > 1$ (see Figure 8.4), and if $D - A' - A$, then $r < 1$. In either case, $\xi_{D,r}(A) = A'$ and $\xi_{D,r}(B) = B'$. If $A - D - A'$, let $E = \xi_{D,r}(A)$ and $F = \xi_{D,r}(B)$. Then by definition, E is the unique point on $\overrightarrow{DA}$ such that $DE = rDA$ and F is the unique point on $\overrightarrow{DB}$ such that $DF = rDB$ (see Figure 8.5). But $DA' = rDA = DE$ and $DB' = rDB = DF$ so that D is the midpoint of $\overline{EA'}$ and $\overline{FB'}$ and it follows that $(\varphi_D \circ \xi_{D,r})(A) = \varphi_D(E) = A'$ and $(\varphi_D \circ \xi_{D,r})(B) = \varphi_D(F) = B'$. Thus in every case, $\delta_{D,r}(A) = A'$ and $\delta_{D,r}(B) = B'$. Therefore $\alpha = \delta_{D,r}$ by the uniqueness in Theorem 264. ■

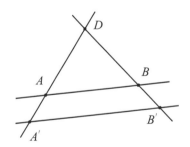

Figure 8.4. Case $r > 1$.

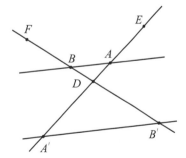

Figure 8.5. Case $r < 1$.

Exercises

1. Let C be a point and let $r > 0$. Prove that the dilation $\varphi_C \circ \xi_{C,r}$ is a similarity of ratio r.

2. (The fixed point of a non-isometric dilation). Let α be a non-isometric dilation, let A and B be distinct points, and let $A' = \alpha(A)$ and $B' = \alpha(B)$. If $A' \neq A$, $B' \neq B$, and $\overleftrightarrow{AA'} \neq \overleftrightarrow{BB'}$, prove that $\overleftrightarrow{AA'} \nparallel \overleftrightarrow{BB'}$ and $\alpha(\overleftrightarrow{AA'} \cap \overleftrightarrow{BB'}) = \overleftrightarrow{AA'} \cap \overleftrightarrow{BB'}$.

3. Let $A = \begin{bmatrix} 0 \\ 0 \end{bmatrix}$, $B = \begin{bmatrix} 1 \\ 1 \end{bmatrix}$, $A' = \begin{bmatrix} 0 \\ 2 \end{bmatrix}$, $B' = \begin{bmatrix} 2 \\ 4 \end{bmatrix}$. Identify the (unique) dilatation α such that $\alpha(A) = A'$ and $\alpha(B) = B'$ as a translation, stretch, or dilation. Determine the ratio of similarity and any fixed points.

8.3 Classification of Similarities and the Similarity Recognition Problem

In this section we prove the premier result in this chapter – the Classification Theorem for Similarities – which asserts that every similarity is either an isometry, a stretch, a stretch reflection, or a stretch rotation. We begin our discussion with the analog Theorem 174 for similar triangles:

Theorem 267 *Triangles $\triangle PQR$ and $\triangle ABC$ are similar if and only if there is a similarity α such that $\triangle ABC = \alpha(\triangle PQR)$. Furthermore, α is unique if and only if $\triangle PQR$ is scalene.*

Proof. ($\Rightarrow$) If $\triangle PQR \sim \triangle ABC$, choose a correspondence of vertices $P \leftrightarrow A$, $Q \leftrightarrow B$, $R \leftrightarrow C$. By Corollary 256 (the Three Points Theorem for Similarities), it is sufficient to define a similarity α with the required property on P, Q, and R. Let $r = AB/PQ$, let $D = \xi_{P,r}(Q)$, and let $E = \xi_{P,r}(R)$. Then $AB = rPQ = PD$, $AC = rPR = PE$, and $\angle DPE = \angle QPR \cong \angle BAC$ are congruent corresponding angles, and $\triangle PDE \cong \triangle ABC$ (SAS). By Theorem 174, there is a unique isometry β such that $\beta(P) = A$, $\beta(D) = B$, and $\beta(E) = C$. Let $\alpha = \beta \circ \xi_{P,r}$. Then

$$\alpha(P) = (\beta \circ \xi_{P,r})(P) = \beta(P) = A,$$

$$\alpha(Q) = (\beta \circ \xi_{P,r})(Q) = \beta(D) = B,$$

$$\alpha(R) = (\beta \circ \xi_{P,r})(R) = \beta(E) = C,$$

and α is a similarity with the required property. Furthermore, if $\triangle PQR$ is not scalene, it is isosceles, and there are at least two different correspondences between the vertices of $\triangle PQR$ and $\triangle ABC$, each of which determines a distinct similarity with the required property. Therefore α is not unique.

($\Leftarrow$) Suppose that α is a similarity of ratio r such that $A = \alpha(P)$, $B = \alpha(Q)$, and $C = \alpha(R)$. Then by definition, $AB = rPQ$, $BC = rQR$, and $CA = rRP$ so that

$$r = \frac{PQ}{AB} = \frac{QR}{BC} = \frac{RP}{CA},$$

and $\triangle PQR \sim \triangle ABC$ by Theorem 65 (Similar Triangles Theorem). Furthermore, if $\triangle PQR$ is scalene, there is a unique correspondence between the vertices of $\triangle PQR$ and $\triangle ABC$, in which case α is unique by the Three Points Theorem for Similarities. ■

Definition 268 *Two plane figures F_1 and F_2 are **similar** if and only if there is a similarity α such that $F_2 = \alpha(F_1)$.*

The proof of Theorem 267 seems to suggest that a similarity is a stretch about some point P followed by an isometry. This is true and very important.

Theorem 269 *If α is a similarity of ratio r and C is any point, there exists an isometry β such that $\alpha = \beta \circ \xi_{C,r}$.*

Proof. Given a similarity α of ratio r, arbitrarily choose a point C. Note that $\beta = \alpha \circ \xi_{C,r}^{-1}$ is an isometry since $\xi_{C,r}^{-1}$ has ratio $\frac{1}{r}$ and the composition $\alpha \circ \xi_{C,r}^{-1}$ has ratio $r \cdot \frac{1}{r} = 1$ by Exercise 8.1.3. Therefore $\alpha = \beta \circ \xi_{C,r}$. ■

In fact, Theorem 269 is so important that we give certain compositions of an isometry and a stretch special names.

Definition 270 *A **stretch rotation** is a non-trivial stretch about a point C followed by a non-trivial rotation about C. A **stretch reflection** is a non-trivial stretch about a point C followed by a reflection in some line through C.*

Example 271 A dilation of the form $\varphi_C \circ \xi_{C,r}$ is a stretch rotation.

Non-isometric similarities have the following important property:

Theorem 272 *Every non-isometric similarity has a fixed point.*

Proof. Let α be a non-isometric similarity. If α is dilatation, it is a dilation by Theorem 266 and has a fixed point. So assume that α is not a dilatation. Let l be a line that cuts its image $l' = \alpha(l)$ at the point $A = l \cap l'$ and let $A' = \alpha(A)$. If $A' = A$, then α has a fixed point and we're done. So assume $A' \neq A$. Then A' is on l' and off l. Let m be the line through A' parallel to l. Since A' is off l, the parallels l and m are distinct. Let $m' = \alpha(m)$.
Claim: Lines l' and m' are distinct parallels. If not, $l' \nparallel m'$ or $l' = m'$, and in either case there exists a point $Q' \in l' \cap m' = \alpha(l) \cap \alpha(m)$ and points Q_1 on l and Q_2 on m such that $Q' = \alpha(Q_1) = \alpha(Q_2)$. But l and m are distinct parallels, hence $Q_1 \neq Q_2$ and α is not injective, which is a contradiction.
Let $B = m \cap m'$ and let $B' = \alpha(B)$. If $B' = B$, then α has a fixed point and we're done. So assume $B' \neq B$. Then B' is on m' and $B' \neq A'$ since A' is on l'. Note that $\overleftrightarrow{AA'} = l' \neq m' = \overleftrightarrow{BB'}$ and $\overleftrightarrow{AA'} \parallel \overleftrightarrow{BB'}$ by the claim. Since α is not an isometry, $\square ABB'A'$ is not a parallelogram and $\overleftrightarrow{AB} \nparallel \overleftrightarrow{A'B'}$. Let $P = \overleftrightarrow{AB} \cap \overleftrightarrow{A'B'}$ and let $P' = \alpha(P)$. If $P' = P$, the proof is complete. We consider three cases:

<u>Case 1</u>: $A - P - B$. Then $A' - P - B'$ since $\overleftrightarrow{AA'} \parallel \overleftrightarrow{BB'}$ (see Figure 8.6) and $A' - P' - B'$ since α preserves betweenness (Exercise 8.1.8). Thus A', P, P', and B' are collinear. Since $\overleftrightarrow{AA'} \parallel \overleftrightarrow{BB'}$, $\angle PAA' \cong \angle PBB'$ and $\angle PA'A \cong \angle PB'B$ by the Alternate Interior Angles Theorem so that $\triangle APA' \sim \triangle BPB'$ (AA). Thus $AP/BP = A'P/B'P$ by Theorem 65 (Similar Triangles Theorem). Let r be the ratio of the similarity α; then $A'P' = rAP$ and $B'P' = rBP$ so that

$$\frac{A'P'}{B'P'} = \frac{rAP}{rBP} = \frac{AP}{BP} = \frac{A'P}{B'P}.$$

Since $A'P' = A'B' - B'P'$ and $A'P = A'B' - B'P$, direct substitution gives

$$\frac{A'B' - B'P'}{B'P'} = \frac{A'B' - B'P}{B'P}.$$

Then $B'P' = B'P$ by algebra, and $P = P'$ as claimed.

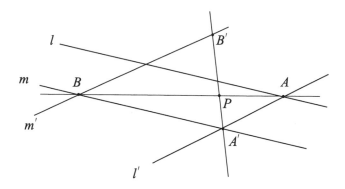

Figure 8.6. Case 1: $A' - P - B'$.

<u>Cases 2 and 3</u>: The proofs assuming $A - B - P$ and $P - A - B$ are similar and left as exercises for the reader. ∎

The table is set for our premier result:

Theorem 273 (Classification of Similarities) *A similarity is exactly one of the following: an isometry, a stretch, a stretch rotation, or a stretch reflection.*

Proof. Consider a non-isometric similarity α, which has a fixed point C by Theorem 272. By Theorem 269, there is an isometry β and a stretch ξ about C such that $\alpha = \beta \circ \xi$. Then $\alpha \circ \xi^{-1} = \beta$ and ξ^{-1} is also a stretch about C. Hence

$$\beta(C) = (\alpha \circ \xi^{-1})(C) = \alpha(C) = C.$$

Since C is a fixed point, β is either the identity, a rotation about C, or a reflection in some line passing through C. Hence α is either a stretch, a stretch rotation, or a stretch reflection. Proof of the fact that α is exactly one of these three is left to the reader. ∎

Theorem 272 tells us that a stretch rotation or a stretch reflection α has a fixed point, so let's construct it. If α is a dilation, the fixed point was constructed in Exercise 8.2.2. So assume that α is not a dilation. By Exercise 8.3.8, there exist non-degenerate triangles $\triangle PQR$ and $\triangle P'Q'R'$ such that $P' = \alpha(P)$, $Q' = \alpha(Q)$, $R' = \alpha(R)$, $\overleftrightarrow{PQ} \nparallel \overleftrightarrow{P'Q'}$, and $\overleftrightarrow{QR} \nparallel \overleftrightarrow{Q'R'}$.

Algorithm 274 *Let α be a non-isometric similarity that is not a dilation. To construct the fixed point of α, choose triangles $\triangle PQR$ and $\triangle P'Q'R'$ that satisfy the conditions of Exercise 8.3.8 and proceed as follows:*

1. *Construct line n through R parallel to $m = \overleftrightarrow{PQ}$.*

2. *Construct line n' through R' parallel to $m' = \overleftrightarrow{P'Q'}$.*

3. *Let $D = m \cap m'$ and $E = n \cap n'$.*

4. *Construct line $a = \overleftrightarrow{DE}$ (see Figure 8.7).*

5. *Interchange P and R and repeat steps 1–4 to construct line b.*

6. *Then $C = a \cap b$ is the fixed point of α.*

Proof. The proof is left to the reader (see Exercises 8.3.8, 8.3.9, and 8.3.10). ∎

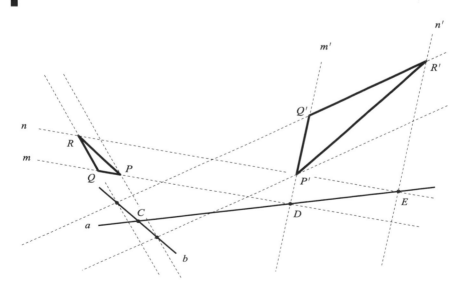

Figure 8.7. Constructing the fixed point of a stretch rotation.

A quite different construction of the fixed point was given by Y. Nishiyama in 2009 [22].

Let α be a similarity of ratio r, let A, B, and C be non-collinear points, let $A' = \alpha(A)$, $B' = \alpha(B)$, and $C' = \alpha(C)$. Recall that α preserves angle measurement up to sign, i.e., $m\angle A'B'C' = \pm m\angle ABC$ (Exercise 7.1.7).

Definition 275 *A similarity α of ratio r is* **direct** *if for all non-collinear points A, B, and C, and $A' = \alpha(A)$, $B' = \alpha(B)$, and $C' = \alpha(C)$, $m\angle A'B'C' = m\angle ABC$; α is* **opposite** *if $m\angle A'B'C' = -m\angle ABC$.*

Proposition 276 *Even isometries, stretches, and stretch rotations are direct similarities; odd isometries and stretch reflections are opposite similarities.*

Proof. First observe that a stretch about the origin O is direct. Let $A = \begin{bmatrix} a_1 \\ a_2 \end{bmatrix}$, $B = \begin{bmatrix} b_1 \\ b_2 \end{bmatrix}$, and $C = \begin{bmatrix} c_1 \\ c_2 \end{bmatrix}$ be non-collinear points. Then $A' = \xi_{O,r}(A) = \begin{bmatrix} ra_1 \\ ra_2 \end{bmatrix}$, $B' = \xi_{O,r}(B) = \begin{bmatrix} rb_1 \\ rb_2 \end{bmatrix}$, and $C' = \xi_{O,r}(C) = \begin{bmatrix} rc_1 \\ rc_2 \end{bmatrix}$, so that

$$\mathbf{B'A'} = r\begin{bmatrix} a_1 - b_1 \\ a_2 - b_2 \end{bmatrix} = r\mathbf{BA} \text{ and } \mathbf{B'C'} = r\begin{bmatrix} c_1 - b_1 \\ c_2 - b_2 \end{bmatrix} = r\mathbf{BC}.$$

Thus

$$\det\left[\mathbf{B'A'} \mid \mathbf{B'C'}\right] = \det\left[r\mathbf{BA} \mid r\mathbf{BC}\right] = r^2 \det\left[\mathbf{BA} \mid \mathbf{BC}\right].$$

Since $\det\left[\mathbf{B'A'} \mid \mathbf{B'C'}\right]$ and $\det\left[\mathbf{BA} \mid \mathbf{BC}\right]$ have the same sign, $m\angle A'B'C' = m\angle ABC$ by Theorem 185. Thus $\xi_{O,r}$ is direct by definition.

A general stretch $\xi_{C,r} = \tau_{\mathbf{OP}} \circ \xi_{O,r} \circ \tau_{\mathbf{OP}}^{-1}$ is direct since translations are direct by Theorem 191. In light of Theorems 191 and 273, even isometries, stretches, and stretch rotations are direct similarities; odd isometries and stretch reflections are opposite similarities. ∎

By the Classification of Similarities (Theorem 273), two similar triangles are related by a similarity of exactly one of the following types: an isometry, a stretch, a stretch rotation, or a stretch reflection. Thus Proposition 276 solves the *Similarity Recognition Problem* – the type of similarity relating a given pair of similar triangles can be recognized at a glance.

We conclude the section by deriving the equations of a similarity.

Theorem 277 *A direct similarity has equations of form*

$$\begin{cases} x' = & ax - by + c \\ y' = & bx + ay + d, \end{cases} \quad a^2 + b^2 > 0;$$

an opposite similarity has equations of form

$$\begin{cases} x' = & ax - by + c \\ y' = & -bx - ay + d, \end{cases} \quad a^2 + b^2 > 0.$$

Conversely, a transformation with equations of either form is a similarity.

Proof. One can easily check that a direct isometry has equations

$$\begin{cases} x' = & ax - by + c \\ y' = & bx + ay + d, \end{cases} \quad a^2 + b^2 = 1;$$

and an opposite isometry has equations

$$\begin{cases} x' = & ax - by + c \\ y' = & -bx - ay + d, \end{cases} \quad a^2 + b^2 = 1.$$

By Theorem 269, every similarity of ratio $r > 0$ can be expressed as a composition of an isometry and $\xi_{O,r}$. Note that $\xi_{O,r}$ is direct and has equations

$$\begin{cases} x' = & rx \\ y' = & ry, \end{cases} \quad r > 0.$$

Composing the equations of a direct isometry with those of $\xi_{O,r}$ to obtain the equations of a direct similarity

$$\begin{cases} x' = & (ra)x - (rb)y + c \\ y' = & (rb)x + (ra)y + d, \end{cases} \quad (ra)^2 + (rb)^2 > 0,$$

which has the desired form. Similarly, composing the equations of an opposite isometry with those of $\xi_{O,r}$ to obtain the equations of an opposite similarity

$$\begin{cases} x' = & (ra)x - (rb)y + c \\ y' = & -(rb)x - (ra)y + d, \end{cases} \quad (ra)^2 + (rb)^2 > 0,$$

which also has the desired form.

The proof of the converse is left to the reader. ∎

Exercises

1. Which points and lines are fixed by a stretch rotation?

2. Which points and lines are fixed by a stretch reflection?

3. Consider an equilateral triangle $\triangle ABC$ and the line $l = \overleftrightarrow{BC}$. Find all points and lines fixed by the similarity $\sigma_l \circ \xi_{A,2}$.

4. The six triangles in the diagram below are similar. Name the similarity that maps $\triangle ABC$ to the five triangles (a) through (e) and state whether the similarity is direct or opposite. If the similarity is a rotation or a stretch rotation, determine the rotation angle.

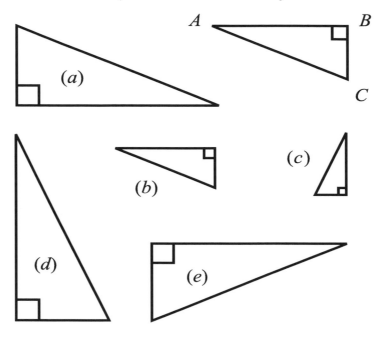

5. A dilation with center P and ratio r has equations $x' = -2x + 3$ and $y' = -2y - 4$. Find P and r.

6. Let α be a similarity such that $\alpha\left(\begin{bmatrix}0\\0\end{bmatrix}\right) = \begin{bmatrix}1\\0\end{bmatrix}$, $\alpha\left(\begin{bmatrix}1\\0\end{bmatrix}\right) = \begin{bmatrix}2\\2\end{bmatrix}$, and $\alpha\left(\begin{bmatrix}2\\2\end{bmatrix}\right) = \begin{bmatrix}-1\\6\end{bmatrix}$.

    (a) Find the equations of α.

    (b) Find $\alpha\left(\begin{bmatrix}-1\\6\end{bmatrix}\right)$.

7. The two triangles in the diagram below are related by a stretch reflection α. Following Algorithm 274, construct the fixed point C of α. Having constructed C, also construct the axis of reflection.

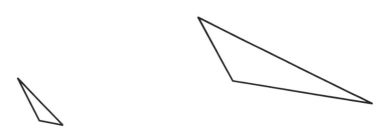

8. Let α be a non-isometric similarity that is not a dilation. Prove that there exist non-degenerate triangles $\triangle PQR$ and $\triangle P'Q'R'$ such that $P' = \alpha(P)$, $Q' = \alpha(Q)$, $R' = \alpha(R)$, $\overleftrightarrow{PQ} \nparallel \overleftrightarrow{P'Q'}$, and $\overleftrightarrow{QR} \nparallel \overleftrightarrow{Q'R'}$.

9. Let α be a non-isometric similarity and let C be the fixed point of α given by Theorem 272. Let l and m be distinct parallel lines off C, let $l' = \alpha(l)$, and let $m' = \alpha(m)$. If $l \cap l' = D$ and $m \cap m' = E$, prove that C, D, and E are collinear.

10. (The fixed-point of a non-isometric similarity that is not a dilation). Let α be a non-isometric similarity that is not a dilation and let C be the fixed point of α given by Theorem 272. Given triangles $\triangle PQR$ and $\triangle P'Q'R'$ that satisfy the conditions of Exercise 8.3.8, let n be the line through R parallel to $m = \overleftrightarrow{PQ}$ and let n' be the line through R' parallel to $m' = \overleftrightarrow{P'Q'}$. Let $D = m \cap m'$ and $E = n \cap n'$. Similarly, let q be the line through P parallel to $p = \overleftrightarrow{QR}$ and let q' be the line through P' parallel to $p' = \overleftrightarrow{Q'R'}$. Let $F = p \cap p'$ and $G = q \cap q'$. Apply Exercise 8.3.9 to prove that $C = \overleftrightarrow{DE} \cap \overleftrightarrow{FG}$.

11. Complete the proof of Theorem 273: Prove that the sets $\mathcal{I} = \{\text{isometries}\}$, $\mathcal{J} = \{\text{non-trivial stretches}\}$, $\mathcal{K} = \{\text{stretch rotations}\}$, and $\mathcal{L} = \{\text{stretch reflections}\}$ are mutually disjoint.

12. Prove that the set of all *direct* similarities forms a group under composition of functions.

13. Which group properties fail for the set of opposite similarities?

14. Complete the proof of Theorem 272:

    (a) If $A - B - P$, prove that $P = P'$.
    (b) If $P - A - B$, prove that $P = P'$.

15. Prove the second half of Theorem 277: A transformation with equations of either indicated form is a similarity.

8.4 Conjugation and Similarity Symmetry

In this section we investigate the geometry of conjugation on the set $\mathcal{S}$ of all similarities and observe that conjugation preserves similarity type. We conclude with a brief discussion of similarity symmetry type.

Consider the conjugation of an isometry by a similarity.

Theorem 278 *The conjugation of an isometry by a similarity is an isometry. More precisely,*

a. *The conjugation of a translation by a similarity is a translation. In fact, if α is a similarity and $\mathbf{v}$ is a vector, then $\alpha \circ \tau_{\mathbf{v}} \circ \alpha^{-1} = \tau_{\alpha(\mathbf{v})}$.*

b. *The conjugation of a rotation by a similarity is a rotation. In fact, if α is a similarity, C is a point, and $\Theta \in \mathbb{R}$, then*

$$\alpha \circ \rho_{C,\Theta} \circ \alpha^{-1} = \begin{cases} \rho_{\alpha(C),\Theta} & \text{if } \alpha \text{ is direct,} \\ \rho_{\alpha(C),-\Theta} & \text{if } \alpha \text{ is opposite.} \end{cases}$$

c. *The conjugation of a reflection by a similarity is a reflection. In fact, if α is a similarity and m is a line, then $\alpha \circ \sigma_m \circ \alpha^{-1} = \sigma_{\alpha(m)}$.*

d. *The conjugation of a glide reflection by a similarity is a glide reflection. In fact, if c is a line and $\mathbf{v}$ is a non-zero vector such that $\tau_{\mathbf{v}}(c) = c$, then $\alpha \circ \gamma_{c,\mathbf{v}} \circ \alpha^{-1} = \gamma_{\alpha(c),\alpha(\mathbf{v})}$.*

Proof. Let α be a similarity of ratio r, and let β be an isometry. The $\alpha \circ \beta \circ \alpha^{-1}$ is an isometry since it is a similarity of ratio $r \cdot 1 \cdot \frac{1}{r} = 1$.

(a) Suppose β is a translation $\tau_{\mathbf{v}}$. Then $\alpha \circ \tau_{\mathbf{v}} \circ \alpha^{-1}$ is direct since $\tau_{\mathbf{v}}$ is direct, and α and α^{-1} are both direct or both opposite. Thus $\alpha \circ \tau_{\mathbf{v}} \circ \alpha^{-1}$ is a translation or a rotation by Theorem 191. If $\mathbf{v} = \mathbf{0}$, then $\alpha \circ \tau_{\mathbf{v}} \circ \alpha^{-1} = \alpha \circ \iota \circ \alpha^{-1} = \iota$ is a translation. If $\mathbf{v} \neq \mathbf{0}$, then $\tau_{\mathbf{v}}$ has no fixed points. Suppose $\alpha \circ \tau_{\mathbf{v}} \circ \alpha^{-1}$ has a fixed point P. Then $(\alpha \circ \tau_{\mathbf{v}} \circ \alpha^{-1})(P) = P$, and applying α^{-1} to both sides gives $\tau_{\mathbf{v}}(\alpha^{-1}(P)) = \alpha^{-1}(P)$. Hence $\alpha^{-1}(P)$ is a point fixed by $\tau_{\mathbf{v}}$, which is a contradiction. Since $\alpha \circ \tau_{\mathbf{v}} \circ \alpha^{-1}$ has no fixed points, it is not be a rotation and is therefore a translation.

To determine the translation vector of $\alpha \circ \tau_{\mathbf{v}} \circ \alpha^{-1}$, consider points P and $Q = \tau_{\mathbf{v}}(P)$ such that $\mathbf{v} = \mathbf{PQ}$. Then $(\alpha \circ \tau_{\mathbf{v}} \circ \alpha^{-1})(\alpha(P)) = (\alpha \circ \tau_{\mathbf{v}})(P) = \alpha(Q)$. Since the translation $\alpha \circ \tau_{\mathbf{v}} \circ \alpha^{-1}$ sends $\alpha(P)$ to $\alpha(Q)$, its translation vector is $\alpha(\mathbf{v})$. Therefore $\alpha \circ \tau_{\mathbf{v}} \circ \alpha^{-1} = \tau_{\alpha(\mathbf{v})}$.

We leave proofs of parts b, c, and d as exercises for the reader. ∎

Theorem 279 *The conjugation of a stretch by a similarity is a stretch. In fact, if α is a similarity and C is a point, then $\alpha \circ \xi_{C,r} \circ \alpha^{-1} = \xi_{\alpha(C),r}$.*

Proof. Let $C = \begin{bmatrix} a \\ b \end{bmatrix}$ and write the equations for $\xi_{C,r}$ in the form

$$\begin{cases} x' = rx - ra + a \\ y' = ry - rb + b. \end{cases}$$

If α is a direct similarity, then by Theorem 277, α has equations of the form

$$\begin{cases} x' = sx - ty + u \\ y' = tx + sy + v, \end{cases} \quad s^2 + t^2 > 0,$$

where s, t, u, and v are constants. Note that $\alpha\left(C\right) = \left[\begin{smallmatrix} sa-tb+u \\ ta+sb+v \end{smallmatrix}\right]$. Let Q be a point and let $\alpha^{-1}\left(Q\right) = \left[\begin{smallmatrix} p \\ q \end{smallmatrix}\right]$. Then $Q = \alpha\left(\left[\begin{smallmatrix} p \\ q \end{smallmatrix}\right]\right) = \left[\begin{smallmatrix} sp-tq+u \\ tp+sq+v \end{smallmatrix}\right]$ and by straightforward algebra we have

$$
\begin{aligned}
\left(\alpha \circ \xi_{C,r} \circ \alpha^{-1}\right)(Q) = \alpha\left(\xi_{C,r}\left(\left[\begin{matrix} p \\ q \end{matrix}\right]\right)\right) &= \alpha\left(\left[\begin{matrix} rp - ra + a \\ rq - rb + b \end{matrix}\right]\right) \\
&= \left[\begin{matrix} s\left(rp - ra + a\right) - t\left(rq - rb + b\right) + u \\ t\left(rp - ra + a\right) + s\left(rq - rb + b\right) + v \end{matrix}\right] \\
&= \left[\begin{matrix} r\left(sp - tq + u\right) - r\left(sa - tb + u\right) + sa - tb + u \\ r\left(tp + sq + v\right) - r\left(ta + sb + v\right) + ta + sb + v \end{matrix}\right] \\
&= \xi_{\alpha(C),r}\left(\left[\begin{matrix} sp - tq + u \\ tp + sq + v \end{matrix}\right]\right) = \xi_{\alpha(C),r}(Q).
\end{aligned}
$$

Thus $\alpha \circ \xi_{C,r} \circ \alpha^{-1} = \xi_{\alpha(C),r}$. If α is an opposite similarity, the proof is similar and left to the reader. ∎

Theorem 280 *The conjugation of a stretch rotation by a similarity is a stretch rotation, i.e., if α is a similarity and β is a stretch rotation, then $\alpha \circ \beta \circ \alpha^{-1}$ is a stretch rotation.*

Proof. Let α be a similarity, and let β be a stretch rotation. By definition, $\beta = \rho_{C,\Theta} \circ \xi_{C,r}$, where C is a point, $\Theta \in \mathbb{R}$, $r > 0$ and $r \neq 1$. Then $\alpha \circ \rho_{C,\Theta} \circ \alpha^{-1} = \rho_{\alpha(C),\pm\Theta}$ by Theorem 278, and $\alpha \circ \xi_{C,r} \circ \alpha^{-1} = \xi_{\alpha(C),r}$ by Theorem 279. Therefore

$$
\alpha \circ \beta \circ \alpha^{-1} = \alpha \circ \rho_{C,\Theta} \circ \xi_{C,r} \circ \alpha^{-1} = \alpha \circ \rho_{C,\Theta} \circ \alpha^{-1} \circ \alpha \circ \xi_{C,r} \circ \alpha^{-1} = \rho_{\alpha(C),\pm\Theta} \circ \xi_{\alpha(C),r}
$$

is a stretch rotation by definition. ∎

Theorem 281 *The conjugation of a stretch reflection by a similarity is a stretch reflection, i.e., if α is a similarity and β is a stretch reflection, then $\alpha \circ \beta \circ \alpha^{-1}$ is a stretch reflection.*

Proof. The proof is similar to that of Theorem 280, and is left to the reader as an exercise. ∎

The results in Theorems 278, 279, 280, and 281 are summarized in Theorem 282 and give us a complete picture of the action of conjugation on the set $\mathcal{S}$ of all similarities.

Theorem 282 *Let α and β be similarities. Then β and $\alpha \circ \beta \circ \alpha^{-1}$ have the same similarity type, i.e., both are isometries, both are stretches, both are stretch rotations, or both are stretch reflections.*

We conclude with some remarks on similarity symmetries, i.e., those similarities that fix a given plane figure. Suppose a stretch $\xi_{C,r}$ fixes a plane figure F. If $\tau_{\mathbf{v}}$ is a similarity symmetry of F, conjugating by $\xi_{C,r}$ and its inverse gives

$$\xi_{C,r} \circ \tau_{\mathbf{v}} \circ \xi_{C,r}^{-1} = \tau_{r\mathbf{v}} \quad \text{and} \quad \xi_{C,r}^{-1} \circ \tau_{\mathbf{v}} \circ \left(\xi_{C,r}^{-1}\right)^{-1} = \tau_{\frac{1}{r}\mathbf{v}}, \qquad (8.2)$$

which are also translational symmetries of F by Theorem 278, part a. Note that one of these conjugates is a translational symmetry of length shorter than $\|\tau_{\mathbf{v}}\|$. Consequently, F has no non-trivial translational symmetry of shortest length. Indeed, the same is true when F is fixed by a stretch rotation or a stretch reflection since conjugating both sides of Equations (8.2) by a rotation about C or a reflection in some line containing C preserves translation length.

So assume that F only has trivial translational symmetry. Then F has no glide reflection symmetry, and the isometric symmetries of F are either point or line symmetries. Also assume that the point symmetries of F have finite order, i.e., each point of symmetry is an n-center for some $n \in \mathbb{N}$.

Now consider a non-isometric similarity symmetry α of F. If C is the fixed point of α, and $P \in F$ is a point of symmetry, then $P = C$ (otherwise $\alpha(P)$ is a point of symmetry distinct from P and there is a non-trivial translational symmetry). Likewise, if m is a line of symmetry, then C is on m (otherwise $m' = \alpha(m)$ is a line of symmetry distinct from m and either $m' \cap m$ is a point of symmetry distinct from C or $m' \parallel m$; but in either case there is a non-trivial translational symmetry). Thus all lines of symmetry are concurrent at C. Consequently, if a plane figure F with point or line symmetry also has non-isometric similarity symmetry, all similarity symmetries of F share the same fixed point.

On the other hand, if α and β are non-isometric similarity symmetries of F with distinct fixed points C and D, then for each $n \in \mathbb{Z}$, $\alpha^n \circ \beta \circ \alpha^{-n}$ is a similarity symmetry with fixed point $D_n = \alpha^n(D)$ and the similarity symmetry group of F is infinitely generated.

Henceforth we limit our considerations to those plane figures whose similarity symmetries share a common fixed point and have finitely generated similarity groups. But before we can proceed, we need a definition.

Let G be a group and let H and K be subgroups of G such that $hk = kh$ for all $h \in H$ and $k \in K$. Then the set $HK = \{hk \mid h \in H, k \in K\}$ is also a subgroup of G, as the reader can easily check.

Definition 283 *Let G be a group and let H and K be subgroups of G. The set HK is the **internal direct product of** H **and** K provided*

1. $H \cap K = \{e\}$, where e is the identity element of G, and

2. $hk = kh$ for all $h \in H$ and $k \in K$.

Since an internal direct product HK is isomorphic to the *direct product* $H \times K = \{(h, k) \mid h \in H, k \in K\}$, where $(h, k)(h', k') := (hh', kk')$, we shall use the symbols HK and $H \times K$ as well as the terms "internal direct product" and "direct product" interchangeably.

Although point and line symmetries commute with stretch symmetries, and point symmetries commute with stretch rotation symmetries, these situations are exceptional. Point symmetries rarely commute with stretch reflections and line symmetries rarely commute with stretch rotations or stretch reflections.

Consider two elements $\sigma_a \circ \xi_r, \sigma_b \circ \xi_s \in \text{Sym}(F)$ (with r or s possibly equal to 1) and let Θ be the measure of an angle from a to b. Then $(\sigma_a \circ \xi_r) \circ (\sigma_b \circ \xi_s) = (\sigma_b \circ \xi_s) \circ (\sigma_a \circ \xi_r)$ if and only if $\rho_{-2\Theta} \circ \xi_{rs} = \rho_{2\Theta} \circ \xi_{rs}$ if and only if $2\Theta \equiv 0, 180$ if and only if $a = b$ or $a \perp b$. Set $r = 1$; then a line symmetry σ_a commutes with a stretch reflection symmetry $\sigma_b \circ \xi_s$ if and only if $a = b$ or $a \perp b$.

Similarly, consider elements $\sigma_a \circ \xi_r, \beta = \rho_\Theta \circ \xi_s \in \text{Sym}(F)$. Then $(\sigma_a \circ \xi_r) \circ (\rho_\Theta \circ \xi_s) = (\rho_\Theta \circ \xi_s) \circ (\sigma_a \circ \xi_r)$ if and only if $\sigma_a \circ \rho_\Theta \circ \xi_{rs} = \rho_\Theta \circ \sigma_a \circ \xi_{rs}$ if and only if $\rho_\Theta = \sigma_a \circ \rho_\Theta \circ \sigma_a$ if and only if $\rho_\Theta = \rho_{-\Theta}$. Set $r = 1$; then a line symmetry σ_a commutes with a stretch rotation symmetry $\rho_\Theta \circ \xi_s$ if and only if ρ_Θ is a halfturn. Set $s = 1$; then a point symmetry ρ_Θ commutes with a stretch reflection symmetry $\sigma_a \circ \xi_r$ if and only if ρ_Θ is a halfturn. This proves the following theorem:

Theorem 284 *Let F be plane figure whose similarity symmetries share a common fixed point. Then* $\text{Sym}(F)$ *splits as the direct product*

$$\text{Sym}(F) \approx \langle \text{isometric symmetries of } F \rangle \times \langle \text{non-isometric symmetries of } F \rangle$$

if and only if any one of the following conditions hold:

1. *The symmetries of F are isometries.*

2. *The only isometric symmetry of F is the identity.*

3. *The only non-trivial isometric symmetry of F is a halfturn.*

4. *The non-isometric symmetries of F are stretches or dilations.*

5. *The symmetries of F are rotations, stretches, or stretch rotations.*

6. *F has one or two line symmetries and at least one stretch reflection symmetry whose axes of stretch reflection are coincident with or perpendicular to the line(s) of symmetry.*

Note that items (1) through (6) are not mutually exclusive; items (3) and (5) can be satisfied simultaneously, for example.

The first step towards classifying the similarity symmetry type of a plane figure F that satisfies one of the conditions in Theorem 284, is to independently identify its isometric and non-isometric symmetries. Note that the subgroup $\langle \text{non-isometric symmetries of } F \rangle \subseteq \text{Sym}(F)$ is abelian whenever (1), (4), (5), or (6) holds. In general, if the subgroup ⟨non-isometric symmetries of F⟩ is abelian and finitely generated, the inner isomorphism class of Sym (F) is a direct product of C_n or D_n with finitely many copies of $\mathbb{Z}$ (non-isometric symmetries have infinite order).

Corollary 285 *Let F be a plane figure whose similarity symmetries share a common fixed point. If* Sym (F) *is finitely generated and the subgroup* ⟨*non-isometric symmetries of F*⟩ $\subseteq$ Sym (F) *is abelian, then*

$$\text{Sym}\,(F) \approx C_n \times \mathbb{Z}^k \ \text{or} \ D_n \times \mathbb{Z}^k \ \text{for some } n \geq 1 \ \text{and } k \geq 0.$$

Corollary 285 falls short of a complete classification of similarity symmetry types even in the restricted setting of the corollary because the factor $\mathbb{Z}^k$ can be generated by multiple non-isometric similarities, which (individually) generate multiple copies of $\mathbb{Z}$ with non-trivial intersection. Of course, if one limits considerations to those plane figures whose similarity symmetry groups admit at most one infinite cyclic factor, the inner isomorphism classes of similarity symmetry types are exactly $C_n, D_n, \mathbb{Z}(i), C_n \times \mathbb{Z}(i)$, and $D_n \times \mathbb{Z}(i)$, $i \in \{1, 2, 3\}$, where $\mathbb{Z}(1)$ is generated by a stretch, $\mathbb{Z}(2)$ is generated by a stretch rotation, and $\mathbb{Z}(3)$ is generated by a stretch reflection (see Figures 8.8, 8.9, and 8.10).

Figure 8.8. A plane figure with similarity symmetry type $\mathbb{Z}(1)$.

Figure 8.9. A plane figure with similarity symmetry type $D_3 \times \mathbb{Z}(1)$.

Figure 8.10. A plane figure with similarity symmetry type $C_3 \times \mathbb{Z}(1)$.

Exercises

1. Let F be the plane figure obtained by replacing each copy of the letter "Y" in Figure 8.8 with the letter "X". Find the similarity symmetry group of F.

2. Find the similarity symmetry group of the following plane figure:

3. Let G be a group and let H and K be subgroups of G.

   (a) Prove that if $hk = kh$ for all $h \in H$ and $k \in K$, then $HK = \{hk \mid h \in H, k \in K\}$ is a subgroup of G.

   (b) Prove that if $H \cap K = \{e\}$ and $hk = kh$ for all $h \in H$ and $k \in K$, then $HK \approx H \times K$.

4. Prove Theorem 278, part b: If α is a similarity, C is a point, and $\Theta \in \mathbb{R}$, then

$$\alpha \circ \rho_{C,\Theta} \circ \alpha^{-1} = \begin{cases} \rho_{\alpha(C),\Theta} & \text{if } \alpha \text{ is direct,} \\ \rho_{\alpha(C),-\Theta} & \text{if } \alpha \text{ is opposite.} \end{cases}$$

5. Prove Theorem 278, part c: If α is a similarity and m is a line, then $\alpha \circ \sigma_m \circ \alpha^{-1} = \sigma_{\alpha(m)}$.

6. Prove Theorem 278, part d: If c is a line and $\mathbf{v}$ is a non-zero vector such that $\tau_{\mathbf{v}}(c) = c$, then $\alpha \circ \gamma_{c,\mathbf{v}} \circ \alpha^{-1} = \gamma_{\alpha(c),\alpha(\mathbf{v})}$.

7. Complete the proof of Theorem 279: If α is an opposite similarity, prove that $\alpha \circ \xi_{C,r} \circ \alpha^{-1} = \xi_{\alpha(C),r}$.

8. Prove Theorem 281: If α is a similarity and β is a stretch reflection, then $\alpha \circ \beta \circ \alpha^{-1}$ is a stretch reflection.

Appendix: Hints and Answers to Selected Exercises

Chapter 1: Axioms of Euclidean Plane Geometry

1.1 The Existence and Incidence Postulates

1. First apply the Existence Postulate, then the Incidence Postulate.

1.2 The Distance and Ruler Postulates

1. (a) Show that (1) g is 1-1; (2) g is onto; and (3) $AB = |g(A) - g(B)|$ for every two points A and B on l.

   (b) Show that (1) h is 1-1; (2) h is onto; and (3) $AB = |h(A) - h(B)|$ for every two points A and B on l.

2. Note that point B lies on $\overrightarrow{AB}$, so if $\overrightarrow{AB} = \overrightarrow{AC}$, then B lies on $\overrightarrow{AC}$.

3. Use the definition of ray $\overrightarrow{AB}$ and Definition 8.

4. Apply the Betweenness Theorem for Points.

5. Suppose that M' is another midpoint of $\overline{AB}$, i.e., $A - M' - B$ and $AM' = M'B$. Then show that $f(M') = f(M)$ where M is the midpoint constructed in the existence proof, and conclude that $M' = M$.

1.3 The Plane Separation Postulate

1. Statement (a) is true, and statement (b) is false.

2. The interior of an angle is the intersection of two half-planes that are convex. Then apply Exercise 1.3.1 (a).

3. Use the definition of ray $\overrightarrow{AB}$ and apply Proposition 17.

4. Show that all points that do not lie on $\overleftrightarrow{BC}$ are on the same side of $\overleftrightarrow{BC}$ as A.

1.4 The Protractor Postulate

1. The existence follows from the existence of midpoints and the Angle Construction Postulate (construct a 90° angle); the uniqueness follows from the uniqueness of midpoints and the uniqueness part of the Angle Construction Postulate.

2. Use the fact that both vertical angles form linear pairs with the same angle.

1.5 The Side-Angle-Side Postulate and the Euclidean Parallel Postulate

1. Triangles $\triangle BAD \cong \triangle CAD$ by SAS.

2. Triangles $\triangle AMB \cong \triangle A'MB'$ by SAS.

3. It is not possible that m and n intersect at some point. Otherwise it violates the Euclidean Parallel Postulate.

4. It is true. When P lies on l, the only line through point P and parallel to l is l itself.

Chapter 2: Theorems of Euclidean Plane Geometry

2.1 The Exterior Angle Theorem

1. Otherwise at least two interior angles are right or obtuse. Then it violates the Exterior Angle Theorem since the exterior angle adjacent to the right or obtuse interior angle will be less than or equal to the other right or obtuse interior angle.

2.2 Triangle Congruence Theorems

1. Consider two cases: A, B, and C are collinear or non-collinear. If they are non-collinear, let D be the point such that $A-B-D$ and $BD = BC$, and apply the Scalene Inequality to $\triangle ACD$.

2. Let B' be the point on $\overrightarrow{CB}$ such that $B'C = EF$. First show $\triangle AB'C \cong \triangle DEF$, and then show $B' = B$ by contradiction.

3. Let A' be the point on $\overrightarrow{CA}$ such that $A'C = DF$. First show $\triangle A'BC \cong \triangle DEF$. Then consider two cases: if $A = A'$, then $\angle BAC$ and $\angle EDF$ are congruent; if $A \neq A'$, then $\angle BAC$ and $\angle EDF$ are supplementary.

4. First use Exercise 2.2.3, and then use Exercise 2.1.1 to exclude the supplementary case.

5. Let M be the midpoint of $\overline{AB}$. If $P = M$, the proof is easy; if $P \neq M$, the first step for each implication is to apply an appropriate triangle congruence theorem to show that $\triangle PAM \cong \triangle PBM$.

6. The goal is to show that $\angle ABC \cong \angle DEF$. Construct point A' on the opposite side of $\overleftrightarrow{BC}$ from A such that $\angle A'BC \cong \angle DEF$ and $A'B = DE$. Then use the fact that $\triangle A'BC \cong \triangle DEF$ to show that $\angle ABC \cong \angle A'BC$. You may find Theorem 48 useful.

7. Let D and E be the feet of the perpendiculars from P to $\overleftrightarrow{AB}$ and $\overleftrightarrow{AC}$, respectively. The first step for each implication is to apply an appropriate triangle congruence theorem to show that $\triangle PAD \cong \triangle PAE$.

2.3 The Alternate Interior Angles Theorem and the Angle Sum Theorem

1. Apply the Vertical Angles Theorem (Theorem 33) and the Alternate Interior Angles Theorem (Theorem 53).

2. Apply the Supplement Postulate and the Alternate Interior Angles Theorem (Theorem 53).

3. Apply the Angle Sum Theorem (Theorem 57) and the Supplement Postulate.

4. (a) Apply the Alternate Interior Angles Theorem (Theorem 53) and an appropriate triangle congruence theorem.

    (b) Use part a.

    (c) Use part a.

    (d) Let M be the point where $\overline{AC}$ and $\overline{BD}$ intersect each other. Use part b, the Alternate Interior Angles Theorem (Theorem 53), and an appropriate triangle congruence theorem to show that $\triangle ABM \cong \triangle CDM$.

5. Use a contradiction proof and apply the Euclidean Parallel Postulate.

2.4 Similar Triangles

1. Since the angle sum of every triangle is $180°$, if two pairs of corresponding angles are congruent, so is the third pair.

2. If $AB = DE$, show that $\triangle ABC \cong \triangle DEF$ so that $\triangle ABC \sim \triangle DEF$; if $AB \neq DE$, without loss of generality, assume that $AB > DE$. Let B' be the point on $\overline{AB}$ such that $AB' = DE$. Let l be the line through point B' and parallel to $\overleftrightarrow{BC}$. By Pasch's Axiom, l intersects $\overline{AC}$ at some point C'. It suffices to show that (1) $\triangle ABC \sim \triangle AB'C'$, and (2) $\triangle AB'C' \cong \triangle DEF$.

Chapter 3: Introduction to Transformations, Isometries, and Similarities

3.1 Transformations

1. α is bijective (both injective and surjective).

   β is neither injective nor surjective.

γ is not injective but is surjective.

δ is bijective (both injective and surjective).

ϵ is neither injective nor surjective.

η is bijective (both injective and surjective).

ρ is injective but not surjective.

σ is bijective (both injective and surjective).

τ is bijective (both injective and surjective).

3.2 Isometries and Similarities

3. (a) $2X - 3Y - 4 = 0$

   (b) $2X - 3Y + 4 = 0$

   (c) $2X + 3Y - 4 = 0$

   (d) $2X + 3Y - 44 = 0$

   (e) $2X + 3Y + 11 = 0$

4. $\alpha\left(\begin{bmatrix} x \\ y \end{bmatrix}\right) = \begin{bmatrix} x \\ y^3 \end{bmatrix}$

5. α is not a collineation.

   β is not a collineation.

   γ is not a collineation.

   δ is a collineation: $\frac{a}{2}X + \frac{b}{3}Y + c = 0$

   ϵ is not a collineation.

   η is a collineation: $\frac{b}{3}X + aY + (-2a + c) = 0.$

   ρ is not a collineation.

   σ is a collineation: $-aX - bY + c = 0.$

   τ is a collineation: $aX + bY + (-2a + 3b + c) = 0.$

6. $X - 10Y - 2 = 0$

14. Apply Exercises 3.1.2, 3.2.11, 3.2.12, 3.2.13, Prop. 81, and Theorem 96.

16. Consider the diameter $\overline{DE}$, and triangles $\triangle DEB$ and $\triangle DEC$.

Chapter 4: Translations, Rotations, and Reflections

4.1 Translations

1. (a) $\begin{bmatrix} 6 \\ -1 \end{bmatrix}$

   (b) $x' = x + 6$, $y' = y - 1$

   (c) $\tau\left(\begin{bmatrix} 0 \\ 0 \end{bmatrix}\right) = \begin{bmatrix} 6 \\ -1 \end{bmatrix}$, $\tau\left(\begin{bmatrix} 3 \\ -7 \end{bmatrix}\right) = \begin{bmatrix} 9 \\ -8 \end{bmatrix}$, $\tau\left(\begin{bmatrix} -5 \\ -2 \end{bmatrix}\right) = \begin{bmatrix} 1 \\ -3 \end{bmatrix}$

   (d) $x = -6$, $y = 1$

2. (a) $\begin{bmatrix} 3 \\ 4 \end{bmatrix}$

   (b) $x' = x + 3$, $y' = y + 4$

   (c) $\tau\left(\begin{bmatrix} 0 \\ 0 \end{bmatrix}\right) = \begin{bmatrix} 3 \\ 4 \end{bmatrix}$, $\tau\left(\begin{bmatrix} 1 \\ 2 \end{bmatrix}\right) = \begin{bmatrix} 4 \\ 6 \end{bmatrix}$, $\tau\left(\begin{bmatrix} -3 \\ -4 \end{bmatrix}\right) = \begin{bmatrix} 0 \\ 0 \end{bmatrix}$

   (d) $x = -3$, $y = -4$

3. (a) $\begin{bmatrix} -7 \\ 6 \end{bmatrix}$

   (b) $x' = x - 7$, $y' = y + 6$

   (c) $\tau_{P,Q}\left(\begin{bmatrix} 3 \\ 6 \end{bmatrix}\right) = \begin{bmatrix} -4 \\ 12 \end{bmatrix}$, $\tau_{P,Q}\left(\begin{bmatrix} 1 \\ 2 \end{bmatrix}\right) = \begin{bmatrix} -6 \\ 8 \end{bmatrix}$, $\tau_{P,Q}\left(\begin{bmatrix} -3 \\ -4 \end{bmatrix}\right) = \begin{bmatrix} -10 \\ 2 \end{bmatrix}$

   (d) $2X + 3Y = 0$

5. For homogeneity, note that an isometry is a collineation and preserves distance and betweenness; for additivity, note that an isometry sends parallelograms to parallelograms.

4.2 Rotations

1. $\rho_{O,30}\left(\begin{bmatrix} 3 \\ 6 \end{bmatrix}\right) = \frac{1}{2}\begin{bmatrix} 3\sqrt{3} - 6 \\ 3 + 6\sqrt{3} \end{bmatrix}$

2. $\rho_{Q,45}\left(\begin{bmatrix} 3 \\ 6 \end{bmatrix}\right) = \frac{1}{2}\begin{bmatrix} 5\sqrt{2} - 6 \\ 7\sqrt{2} + 10 \end{bmatrix}$

3. (a) $(\sqrt{3} - \frac{3}{2})X + (1 + \frac{3\sqrt{3}}{2})Y + 4 = 0$

   (b) $-\frac{\sqrt{2}}{2}X + \frac{5\sqrt{2}}{2}Y + 13 - 14\sqrt{2} = 0$

9. $\frac{1}{2}\begin{bmatrix} 3 \\ -8 \end{bmatrix}$

10. (a) $x' = 4 - x$, $y' = 6 - y$

(b) $\varphi_P \left(\begin{bmatrix} 1 \\ 2 \end{bmatrix} \right) = \begin{bmatrix} 3 \\ 4 \end{bmatrix}$, $\varphi_P \left(\begin{bmatrix} -2 \\ 5 \end{bmatrix} \right) = \begin{bmatrix} 6 \\ 1 \end{bmatrix}$

(c) $5X - Y - 21 = 0$

11. (a) $x' = -6 - x$, $y' = 4 - y$

(b) $\varphi_P \left(\begin{bmatrix} 1 \\ 2 \end{bmatrix} \right) = \begin{bmatrix} -7 \\ 2 \end{bmatrix}$, $\varphi_P \left(\begin{bmatrix} -2 \\ 5 \end{bmatrix} \right) = \begin{bmatrix} -4 \\ -1 \end{bmatrix}$

(c) $5X - Y + 27 = 0$

12. The equations of the composition $\varphi_Q \circ \varphi_P$ are $x'' = x - 16$, $y'' = y + 18$. Hence $\varphi_Q \circ \varphi_P$ is a translation by vector $\begin{bmatrix} -16 \\ 18 \end{bmatrix}$.

13. The equations of the composition $\varphi_Q \circ \varphi_P$ are $x'' = x + 2(c - a)$, $y'' = y + 2(d - b)$. Hence $\varphi_Q \circ \varphi_P$ is a translation by vector $\begin{bmatrix} 2(c - a) \\ 2(d - b) \end{bmatrix}$.

4.3 Reflections

1. (a) WOW, AVA, OTTO, YAY, AHA

   (b) HIKE, CHICK, BIB, COB,DID

2. **L** or **F**

5. (a) Let $A' = \sigma_r(A)$. The desired location is $C = \overleftrightarrow{A'B} \cap r$.

   (b) $C = \begin{bmatrix} 4 \\ 0 \end{bmatrix}$

   (c) $5\sqrt{5}$

   (d) Use the triangle inequality.

6. (a) Let $A' = \sigma_q(A)$ and let $A'' = \sigma_p(A')$. Let $C = \overleftrightarrow{BA''} \cap p$ and $D = \overleftrightarrow{CA'} \cap q$. The desired path is $\overline{AD} \cup \overline{DC} \cup \overline{CB}$.

   (b) $D = \begin{bmatrix} 4 \\ 5 \end{bmatrix}$ and $C = \begin{bmatrix} 14 \\ 0 \end{bmatrix}$.

   (c) $8\sqrt{5}$

10. 97 cm

11. $\sigma_l \left(\begin{bmatrix} x \\ y \end{bmatrix} \right) = \frac{1}{5} \begin{bmatrix} 3x - 4y + 12 \\ -4x - 3y + 24 \end{bmatrix}$; $\sigma_l \left(\begin{bmatrix} -5 \\ 3 \end{bmatrix} \right) = \begin{bmatrix} -3 \\ 7 \end{bmatrix}$

12. See the table below.

Equation of l	P	$\sigma_l(P)$	Equation of l	P	$\sigma_l(P)$
$X = 0$	$\begin{bmatrix} x \\ y \end{bmatrix}$	$\begin{bmatrix} -x \\ y \end{bmatrix}$	$Y = -3$	$\begin{bmatrix} x \\ y \end{bmatrix}$	$\begin{bmatrix} x \\ -6-y \end{bmatrix}$
$Y = 0$	$\begin{bmatrix} x \\ -y \end{bmatrix}$	$\begin{bmatrix} x \\ y \end{bmatrix}$	$X = -\frac{3}{2}$	$\begin{bmatrix} 5 \\ 3 \end{bmatrix}$	$\begin{bmatrix} -8 \\ 3 \end{bmatrix}$
$Y = X$	$\begin{bmatrix} 3 \\ 2 \end{bmatrix}$	$\begin{bmatrix} 2 \\ 3 \end{bmatrix}$	$Y = -X$	$\begin{bmatrix} 0 \\ 3 \end{bmatrix}$	$\begin{bmatrix} -3 \\ 0 \end{bmatrix}$
$Y = X$	$\begin{bmatrix} x \\ y \end{bmatrix}$	$\begin{bmatrix} y \\ x \end{bmatrix}$	$Y = -X$	$\begin{bmatrix} -y \\ -x \end{bmatrix}$	$\begin{bmatrix} x \\ y \end{bmatrix}$
$X = 2$	$\begin{bmatrix} -2 \\ 3 \end{bmatrix}$	$\begin{bmatrix} 6 \\ 3 \end{bmatrix}$	$Y = 2X$	$\begin{bmatrix} 0 \\ 5 \end{bmatrix}$	$\begin{bmatrix} 4 \\ 3 \end{bmatrix}$
$Y = -3$	$\begin{bmatrix} -4 \\ -1 \end{bmatrix}$	$\begin{bmatrix} -4 \\ -5 \end{bmatrix}$	$Y = X + 5$	$\begin{bmatrix} x \\ y \end{bmatrix}$	$\begin{bmatrix} y-5 \\ x+5 \end{bmatrix}$

13. (a) $X + Y = 0$

    (b) $X + Y - 10 = 0$

14. $\sigma_l\left(\begin{bmatrix} 0 \\ 0 \end{bmatrix}\right) = \begin{bmatrix} 4 \\ -2 \end{bmatrix}$, $\sigma_l\left(\begin{bmatrix} 1 \\ 3 \end{bmatrix}\right) = \begin{bmatrix} 1 \\ -3 \end{bmatrix}$, $\sigma_l\left(\begin{bmatrix} -2 \\ 1 \end{bmatrix}\right) = \begin{bmatrix} 6 \\ -3 \end{bmatrix}$,

    $\sigma_l\left(\begin{bmatrix} 2 \\ 4 \end{bmatrix}\right) = \begin{bmatrix} 6 \\ 2 \end{bmatrix}$

15. $\sigma_m\left(\begin{bmatrix} 0 \\ 0 \end{bmatrix}\right) = \frac{1}{5}\begin{bmatrix} -6 \\ 12 \end{bmatrix}$, $\sigma_m\left(\begin{bmatrix} 4 \\ -1 \end{bmatrix}\right) = \frac{1}{5}\begin{bmatrix} 2 \\ 31 \end{bmatrix}$, $\sigma_m\left(\begin{bmatrix} -3 \\ 5 \end{bmatrix}\right) = \begin{bmatrix} 1 \\ -3 \end{bmatrix}$,

    $\sigma_m\left(\begin{bmatrix} 3 \\ 6 \end{bmatrix}\right) = \frac{1}{5}\begin{bmatrix} 27 \\ 6 \end{bmatrix}$

16. $18X - Y + 44 = 0$

17. (a) $\sigma_l : x' = -y + 2,\ y' = -x + 2$; $\sigma_m : x'' = -y' - 8,\ y'' = -x' - 8$

    $\Rightarrow \sigma_m \circ \sigma_l : x'' = x - 10,\ y'' = y - 10$.

    (b) $\|\mathbf{v}\| = 10\sqrt{2}$. The distance between l and m is $5\sqrt{2}$.

18. (a) $\sigma_l : x' = -y + 2,\ y' = -x + 2$; $\sigma_m : x'' = y' - 8,\ y'' = x' + 8$

    $\Rightarrow \sigma_m \circ \sigma_l : x'' = -6 - x,\ y'' = 10 - y$.

    (b) $C = \begin{bmatrix} -3 \\ 5 \end{bmatrix}$

    (c) $P = \begin{bmatrix} -3 \\ 5 \end{bmatrix}$. We observe that $C = P$.

21. All real numbers $a = b \neq 0$.

Chapter 5: Compositions of Translations, Rotations, and Reflections

5.2 Rotations as Compositions of Two Reflections

1. (a) $C = \begin{bmatrix} 3 \\ 3 \end{bmatrix}$, $\Theta' = -90$

   (b) $x' = y$, $y' = -x + 6$

   (c) The measures of the angles from l to m are 135 and -45, and Θ' is congruent to twice the measure of each angle (mod 360).

   (d) $\sigma_l : x' = -x + 6$, $y' = y$; $\sigma_m : x'' = y'$, $y'' = x' \Rightarrow \sigma_m \circ \sigma_l : x'' = y$, $y'' = 6 - x$. We observe that the equations of $\sigma_m \circ \sigma_l$ coincide with the equations of $\rho_{C,\Theta}$ in part (b).

2. (a) $C = \begin{bmatrix} -1 \\ 3 \end{bmatrix}$, $\Theta' = 90$

   (b) $x' = -y + 2$, $y' = x + 4$

   (c) The measures of the angles from l to m are 45 and -135, and Θ' is congruent to twice the measure of each angle (mod 360).

   (d) $\sigma_l : x' = -y + 2$, $y' = -x + 2$; $\sigma_m : x'' = x'$, $y'' = -y' + 6$ $\Rightarrow \sigma_m \circ \sigma_l : x'' = -y + 2$, $y'' = x + 4$. We observe that the equations of $\sigma_m \circ \sigma_l$ coincide with the equations of $\rho_{C,\Theta}$ in part (b).

3. (a) $C = \begin{bmatrix} 2 \\ 2 \end{bmatrix}$, $\Theta' = 180$

   (b) $x' = 4 - x$, $y' = 4 - y$

   (c) The measures of the angles from l to m are 90 and -90, and Θ' is congruent to twice the measure of each angle (mod 360).

   (d) $\sigma_l : x' = y$, $y' = x$; $\sigma_m : x'' = 4 - y'$, $y'' = 4 - x' \Rightarrow \sigma_m \circ \sigma_l : x'' = 4 - x$, $y'' = 4 - y$. We observe that the equations of $\sigma_m \circ \sigma_l$ coincide with the equations of $\rho_{C,\Theta}$ in part (b).

4. Any two lines l and m intersecting at the origin with directed angle from l to m of $45°$; e.g., $l : Y = 0$; $m : Y = X$.

5. Any two lines l and m intersecting at $C = \begin{bmatrix} 3 \\ 4 \end{bmatrix}$ with directed angle from l to m of $30°$; e.g., $l : Y = 4$; $m : X - \sqrt{3}Y + 4\sqrt{3} - 3 = 0$.

6. (a) $p : Y = -\frac{1}{2}X$

   (b) $q : Y = \frac{1}{2}X$

5.3 Translations as Compositions of Two Halfturns or Two Reflections

1. $x' = x$, $y' = y + 4$

2. $x' = x - 4$, $y' = y + 4$

3. $l : 8X - 6Y = 0$; $m : 8X - 6Y - 25 = 0$

4. $l : 4X - 2Y = 0$; $m : 4X - 2Y - 15 = 0$

5. (a) $p : Y = 11$

   (b) $q : Y = 7$

7. Right-side-up. The transformation can be thought of as the composition of two halfturns and is therefore a translation.

9. Use Theorem 145 to replace τ_{AB} and τ_{BA} with the appropriate compositions of halfturns.

5.4 The Angle Addition Theorem

1. (a) $l : X - Y = 0$; $m : Y = 0$; $n : X + Y - 2 = 0$

   (b) $D = \begin{bmatrix} 1 \\ 1 \end{bmatrix}$

   (c) $E = \begin{bmatrix} 1 \\ -1 \end{bmatrix}$

2. (a) $l : Y = 0$; $m : X = 0$; $n : X + \sqrt{3}Y - \sqrt{3} = 0$

   (b) $D = \begin{bmatrix} \sqrt{3} \\ 0 \end{bmatrix}$, $\Theta \in 300°$

   (c) $E = \frac{1}{4} \begin{bmatrix} \sqrt{3} \\ 3 \end{bmatrix}$, $\Phi \in 180°$

3. (a) $l : X - 4 = 0$; $m : X + Y - 4 = 0$; $n : (\sqrt{3} - 2)X + Y - 4 = 0$

   (b) $C = \begin{bmatrix} 4 \\ 12 - 4\sqrt{3} \end{bmatrix}$, $\Theta \in 210°$

   (c) $D = \begin{bmatrix} 4\sqrt{3} - 8 \\ 0 \end{bmatrix}$, $\Phi \in 210°$

5.5 Glide Reflections

1. (a) translation or halfturn

   (b) translation or halfturn

   (c) reflection

   (d) glide reflection

   (e) glide reflection

   (f) translation or halfturn

(g) translation or halfturn

(h) translation or halfturn

10. (a) $\sigma_c : \begin{cases} x' = \frac{3}{5}x + \frac{4}{5}y - \frac{6}{5} \\ y' = \frac{4}{5}x - \frac{3}{5}y + \frac{12}{5}, \end{cases}$ $\tau_{PQ} : \begin{cases} x'' = x' + 4 \\ y'' = y' + 2, \end{cases}$

$\Rightarrow \gamma = \tau_{PQ} \circ \sigma_c : \begin{cases} x'' = \frac{3}{5}x + \frac{4}{5}y + \frac{14}{5} \\ y'' = \frac{4}{5}x - \frac{3}{5}y + \frac{22}{5}. \end{cases}$

(b) $\gamma \left(\begin{bmatrix} 1 \\ 2 \end{bmatrix} \right) = \begin{bmatrix} 5 \\ 4 \end{bmatrix}, \gamma \left(\begin{bmatrix} -2 \\ 5 \end{bmatrix} \right) = \frac{1}{5} \begin{bmatrix} 28 \\ -1 \end{bmatrix}, \gamma \left(\begin{bmatrix} -3 \\ -2 \end{bmatrix} \right) = \frac{1}{5} \begin{bmatrix} -3 \\ 16 \end{bmatrix}.$

(c) $\gamma = \tau_{PQ} \circ \sigma_c$ is a glide reflection since line $X - 2Y + 3 = 0$ is parallel to vector $\mathbf{PQ} = \begin{bmatrix} 4 \\ 2 \end{bmatrix}$.

13. The glide vector is $\frac{1}{2}\mathbf{v}$ and the axis is any line c in the direction of $\mathbf{v}$.

Chapter 6: Classification of Isometries

6.1 The Fundamental Theorem and Congruence

1. (a) $\triangle ABC = \sigma_l(\triangle PQR)$, where $l : X = 0$ is the axis of reflection.

(b) $\triangle ABC = (\sigma_m \circ \sigma_l)(\triangle PQR)$, where $l : 4X + 2Y - 45 = 0$ and $m : 2X + Y - 35 = 0$. The composition is a translation by vector $\begin{bmatrix} 10 \\ 5 \end{bmatrix}$.

(c) $\triangle ABC = (\sigma_m \circ \sigma_l)(\triangle PQR)$, where $l : X = 0$ and $m : Y = 0$. The composition is a halfturn centered at $\begin{bmatrix} 0 \\ 0 \end{bmatrix}$.

(d) $\triangle ABC = (\sigma_m \circ \sigma_l)(\triangle PQR)$, where $l : 2Y - 15 = 0$ and $m : X - Y + 15 = 0$. The composition is a rotation of $90°$ centered at $\begin{bmatrix} -15/2 \\ 15/2 \end{bmatrix}$.

Note: $\sigma_l(Q) = B$.

(e) $\triangle ABC = (\sigma_n \circ \sigma_m \circ \sigma_l)(\triangle PQR)$, where $l : 2X - 4Y - 35 = 0$, $m : 2X + Y - 10 = 0$, and $n : Y + 10 = 0$. The composition is a glide reflection with axis $2X - 15 = 0$ and glide vector $\begin{bmatrix} 0 \\ -10 \end{bmatrix}$.

2. (a) $\triangle ABC = \sigma_l(\triangle PQR)$, where $l : X + Y - 28 = 0$ is the axis of reflection.

(b) $\triangle ABC = (\sigma_m \circ \sigma_l)(\triangle PQR)$, where $l : 4X - 3Y + 1 = 0$ and $m : 4X - 3Y - 24 = 0$. The composition is a translation by vector $\begin{bmatrix} 8 \\ -6 \end{bmatrix}$.

(c) $\triangle ABC = (\sigma_m \circ \sigma_l)(\triangle PQR)$, where $l : X = 0$ and $m : Y - 14 = 0$. The composition is a halfturn centered at $\begin{bmatrix} 0 \\ 14 \end{bmatrix}$.

(d) $\triangle ABC = (\sigma_m \circ \sigma_l)(\triangle PQR)$, where $l : X - 4 = 0$ and $m : X + Y - 4 = 0$. The composition is a rotation of $90°$ centered at $\begin{bmatrix} 4 \\ 0 \end{bmatrix}$.

Note: $\sigma_l(R) = C$.

(e) $\triangle ABC = (\sigma_n \circ \sigma_m \circ \sigma_l)(\triangle PQR)$, where $l : 5X + 7Y - 74 = 0$, $m : 7X - 5Y = 0$, and $n : Y = 0$. The composition is a glide reflection with axis $X - 5 = 0$ and glide vector $\begin{bmatrix} 0 \\ -14 \end{bmatrix}$.

Note: $\sigma_l(Q) = B$ and $(\sigma_m \circ \sigma_l)(P) = A$.

6.2 Classification of Isometries

1. $l : X - Y = 0$, $m : X - Y + 2 = 0$; $\sigma_m \circ \sigma_l$ is a translation.

2. $l : X + Y = 0$, $m : X - Y + 1 = 0$; $\sigma_m \circ \sigma_l$ is a rotation (halfturn).

3. $l : X + Y = 0$, $m : X + Y = 0$; $\sigma_m \circ \sigma_l$ is the identity.

4. $l : X + Y = 0$, $m : X + Y - 4 = 0$; $\sigma_m \circ \sigma_l$ is a translation.

5. $l : X + 3Y = 0$, $m : X - 2Y + 5 = 0$; $\sigma_m \circ \sigma_l$ is a rotation.

6. $l : X + 3Y = 0$, $m : X + 3Y - 4 = 0$; $\sigma_m \circ \sigma_l$ is a translation.

8. Isometric dilatations are halfturns or translations. Rotations other than halfturns are not dilatations. Reflections and glide reflections are not dilatations since the image of a line a under reflection in line l is not parallel to a unless l and a are parallel or perpendicular. By the Classification Theorem for Isometries, this exhausts all possibilities.

6.3 Orientation and the Isometry Recognition Problem

1. (a) positive orientation
   (b) negative orientation

2. (a) reverses orientation
   (b) preserves orientation
   (c) preserves orientation
   (d) preserves orientation
   (e) reverses orientation

3. (a) halfturn, direct, 180

   (b) reflection, opposite

   (c) translation, direct

   (d) rotation, direct, 90

   (e) glide reflection, opposite

6.4 The Geometry of Conjugation

1. Line b is the image of line a under a halfturn about point B: $b = \varphi_B(a)$.

2. Point B is the reflection of point A in line b: $B = \sigma_b(A)$.

Chapter 7: Symmetry of Plane Figures

7.2 Symmetry Type

1. $\rho_{120}^0 = \iota$, $\rho_{120}^1 = \rho_{120}$, $\rho_{120}^2 = \rho_{240}$, $\sigma_l^1 = \sigma_l$, $\rho_{120} \circ \sigma_l = \sigma_m$, and $\rho_{120}^2 \circ \sigma_l = \sigma_n$. Therefore $D_3 = \langle \rho_{120}, \sigma_l \rangle$. Furthermore, no single element of D_3 generates the entire group:

$$\iota^n = \iota \qquad \forall n \in \mathbb{Z}$$

$$
\begin{array}{llll}
\rho_{120}^0 = \iota, & \rho_{120}^1 = \rho_{120}, & \rho_{120}^2 = \rho_{240}, & \rho_{120}^3 = \iota, \quad \cdots \\
\rho_{240}^0 = \iota, & \rho_{240}^1 = \rho_{240}, & \rho_{240}^2 = \rho_{120}, & \rho_{240}^3 = \iota, \quad \cdots \\
\sigma_l^0 = \iota, & \sigma_l^1 = \sigma_l, & \sigma_l^2 = \iota, & \cdots \\
\sigma_m^0 = \iota, & \sigma_m^1 = \sigma_m, & \sigma_m^2 = \iota, & \cdots \\
\sigma_n^0 = \iota, & \sigma_n^1 = \sigma_n, & \sigma_n^2 = \iota, & \cdots
\end{array}
$$

2. The Cayley table for D_4:

$\circ$	ι	ρ_{90}	ρ_{180}	ρ_{270}	σ_a	σ_b	σ_c	σ_d
ι	ι	ρ_{90}	ρ_{180}	ρ_{270}	σ_a	σ_b	σ_c	σ_d
ρ_{90}	ρ_{90}	ρ_{180}	ρ_{270}	ι	σ_b	σ_c	σ_d	σ_a
ρ_{180}	ρ_{180}	ρ_{270}	ι	ρ_{90}	σ_c	σ_d	σ_a	σ_b
ρ_{270}	ρ_{270}	ι	ρ_{90}	ρ_{180}	σ_d	σ_a	σ_b	σ_c
σ_a	σ_a	σ_d	σ_c	σ_b	ι	ρ_{270}	ρ_{180}	ρ_{90}
σ_b	σ_b	σ_a	σ_d	σ_c	ρ_{90}	ι	ρ_{270}	ρ_{180}
σ_c	σ_c	σ_b	σ_a	σ_d	ρ_{180}	ρ_{90}	ι	ρ_{270}
σ_d	σ_d	σ_c	σ_b	σ_a	ρ_{270}	ρ_{180}	ρ_{90}	ι

3. (a) C_2 (b) D_2

4. **F, G, J, L, P, Q**, and **R** have symmetry group C_1.

   N, S, and **Z** have symmetry group C_2.

   A, B, C, D, E, K, M, T, U, V, W, and **Y** have symmetry group D_1.

   H, I, and **O** have symmetry group D_2.

   X has symmetry group D_4.

5. $\quad\quad\quad\quad C_6 \quad\quad C_2 \quad\quad D_4$

$\quad\quad\quad\quad\quad D_5 \quad\quad D_8 \quad\quad D_5$

$\quad D_1$

$\quad\quad\quad\quad\quad D_1 \quad\quad D_2 \quad\quad C_5$

$\quad\quad\quad\quad\quad\quad\quad D_1$

6. Consider a square $\square ABCD$ with centroid E and vertices labeled clockwise. The pinwheel constructed in the text (see Figure 7.2) begins by subdividing $\square ABCD$ into four 45-45-90 triangles whose right angles are at E and whose acute angles are at A, B, C, and D. One then cuts along $\overline{AE}$, stopping at its midpoint M. One makes similar cuts along $\overline{BE}, \overline{CE}$, and $\overline{DE}$. With the square positioned so that $\triangle ABE$ is at the top, fold $\triangle ABE$ along the median $\overline{BM}$. Rotate 90 degrees and repeat.

7.3 Rosettes

1. **H, I, N, O** (written as a non-circular oval), **S, X**, and **Z** are rosettes.

2. (a) If $n \geq 2$ is even, the graph of $r = \cos n\theta$ is a rosette with symmetry group D_{2n}. If $n \geq 2$ is odd, the graph of $r = \cos n\theta$ is a rosette with symmetry group D_n.

(b) The graph of $r = \cos \theta$ is a circle centered at $\begin{bmatrix} 1/2 \\ 0 \end{bmatrix}$, which is not a rosette since no point symmetry has a minimum positive rotation angle.

4. First row: D_2, C_5, C_4

Second row: C_3, not a rosette, not a rosette

Third row: C_6, D_3, C_8

Fourth row: C_6, D_6, C_{12}

Fifth row: D_8, C_{14}, C_{16}

7.4 Frieze Patterns

2. F_7

3. The frieze groups F_3 and F_7 are non-abelian since two distinct vertical line symmetries do not commute. The frieze groups F_4 and F_5 are non-abelian since two distinct point symmetries do not commute.

5. (a) F_5 (b) F_4 (c) F_4 (d) F_2 (e) F_6

6. (a) F_4 (b) F_5 (c) F_4 (d) F_2 (e) F_4 (f) F_7 (g) F_6 (h) F_5 (i) F_3 (j) F_3 (k) F_5 (l) F_5 (m) F_6 (n) F_7

7. (a) F_7 (b) F_5 (c) not a frieze (no basic translation) (d) F_5 (e) F_3 (f) F_4 (g) F_6 (h) F_7 (i) F_2 (j) F_2 (k) F_6 (l) F_3 (m) F_1 (n) F_1

7.5 Wallpaper Patterns

1. *p4m*

3. Given non-trivial rotations $\rho_{A,\Theta}, \rho_{B,\Phi} \in \text{Sym}\,(F)$, show that $\rho_{B,\Phi} \circ \rho_{A,\Theta} \circ \rho_{B,-\Phi} \circ \rho_{A,-\Theta} \in \text{Sym}\,(F)$ is a non-trivial translation.

4. Consider two cases: P is or is not a rectangle. When P is not a rectangle, apply Exercise 7.5.3.

5. Set up and solve a system of two linear equations in five unknowns – the number of interior angles in a generating polygon measuring 30, 45, 60, 90, and 120.

6. First row: *p2, p3, p6*

   Second row: *p31m, p4, pgg*

   Third row: *p3m1, cm, pmg*

7. First row: *p4g, p2, p4m*

   Second row: *cm, pgg, p3*

   Third row: *pmm, cmm, p6m*

Chapter 8: Similarity

8.1 Plane Similarities

1. Let E and F be the points at the observer's eye and feet. Let T and B be the points at the top and base of the object. Let M be the surface of the mirror. Then $\triangle EFM$ is similar to $\triangle TBM$. Distances EF, FM, and BM are easily measured and TB can then be determined from the ratio of similarity BM/FM.

2. $r = \frac{5\sqrt{2}}{4}$

3. $P = \frac{1}{2}\begin{bmatrix} -7 \\ 5 \end{bmatrix}$ and $r = 3$.

6. See Exercises 3.1.3, 3.1.4, 3.2.13, and 3.2.15.

8.2 Classification of Dilatations

3. $\alpha = \xi_{C,2}$ is a stretch, where the ratio $r = 2$ and the fixed point $C = \begin{bmatrix} 0 \\ -2 \end{bmatrix}$.

8.3 Classification of Similarities and the Similarity Recognition Problem

1. A stretch rotation fixes its center, a non-stretch dilation fixes every line passing through its center, and other stretch rotations fix no lines.

2. A stretch reflection fixes its center, its axis of reflection l, and the line perpendicular to l passing through its center.

3. Let D be the foot of the perpendicular n from A to $l = \overleftrightarrow{BC}$, let P be the point on $\overline{AD}$ such that $AP = 2PD$, and let m be the line through P and perpendicular to n. Then $\sigma_l \circ \xi_{A,2}$ has a unique fixed point P and two fixed lines m and n.

4. (a) stretch rotation, direct, 180

   (b) stretch, direct

   (c) stretch rotation, direct, 270

   (d) stretch reflection, opposite

   (e) stretch reflection, opposite

5. $P = \frac{1}{3}\begin{bmatrix} 3 \\ -4 \end{bmatrix}$ and $r = 2$.

6. (a) $x' = x - 2y + 1$, $y' = 2x + y$

   (b) $\alpha\left(\begin{bmatrix} -1 \\ 6 \end{bmatrix}\right) = \begin{bmatrix} -12 \\ 4 \end{bmatrix}$

9. Show that there is a dilation $\delta_{C,s}$ such that $\delta_{C,s}(l) = m$ and $\delta_{C,s}(l') = m'$.

8.4 Conjugation and Similarity Symmetry

1. $D_1 \times \mathbb{Z}(1)$

2. $D_6 \times \mathbb{Z}(1)$

Bibliography

1. Armstrong, M. A., *Groups and Symmetry*, Springer-Verlag, New York, 1988.

2. Ball, Philip, *Fearful symmetry: Roger Penrose's tiling*, Prospect Magazine, London, 2013.

3. Baxter, Andrew M. and Umble, Ronald, *Periodic orbits for billiards on an equilateral triangle*, Amer. Math. Monthly., 115(8) (2006) 479–491.

4. Brown, Richard G., *Transformational Geometry*, Dale Seymour Publications, Palo Alto, CA, 1973.

5. Common Core State Standards Initiative [CCSSI], *Common Core State Standards for Mathematics*, National Governors Association Center for Best Practices and the Council of Chief State School Officers, Washington, DC, 2010.

6. Dayoub, Iris Mack and Lott, Johnny W., *Geometry: Constructions and Transformations*, Dale Seymour Publications, Palo Alto, CA, 1977.

7. De Villiers, Michael D., *Rethinking Proof with the Geometers Sketchpad*, Key Curriculum Press, Emeryville, CA, 2003.

8. Descartes, René, *La Géométrie, Third Appendix of Discours de la Méthode*, Leiden, Netherlands, 1637.

9. Dodge, Clayton W., *Euclidean Geometry and Transformations*, Addison-Wesley, Reading, MA, 1972.

10. Emert, John W., Meeks, Kay I., and Nelson, Roger B., *Reflections on a Mira*, The American Mathematical Monthly 101 (1994), 544–549.

11. Eves, Howard, *Return to Mathematical Circles: A Fifth Collection of Mathematical Stories and Anecdotes*, PWS-Kent Publishing Company, Boston, 1988.

12. Eves, Howard, *College Geometry*, Jones and Bartlett, Boston, 1995.

13. Guggenheimer, Heinrich W., *Plane Geometry and Its Groups*, Holden-Day, San Francisco, 1967.

14. Heath, Sir Thomas L., *The Thirteen Books of Euclid's Elements with Introduction and Commentary*, Dover Publications, New York, 1956.

15. Jones, Keith, *Issues in the teaching and learning of geometry. In Linda Haggarty (Ed)*, Aspects of Teaching Secondary Mathematics, Routledge-Falmer, London, (2002) 121–139.

16. Kirby, M. and Umble, R., *Edge tessellations and stamp folding puzzles*, Mathematics Magazine, 84 (2011) 283–289.

17. Kline, Morris, *Mathematics for the Nonmathematician*, Dover Publications, New York, 1985.

18. Lehmann, W.P., *Contemporary Linguistics and Indo-European Studies*, Modern Language Association, 87 (1972) 976–993.

19. Lynch, Mark A. M. and Fraser, Sylvia J., *Similarity Group, Geometry and Symmetry*, International Journal of Mathematical Education in Science and Technology, 30 (1999) 823–831.

20. Martin, George E., *Transformation Geometry: An Introduction to Symmetry*, Springer-Verlag, New York, 1982.

21. National Council of Teachers of Mathematics, *Principles and Standards for School Mathematics*, NCTM, 2000.

22. Nishiyama, Y., *Fixed Points in Similarity Transformations*, International Journal of Pure and Applied Mathematics, 56 (2009) 429–438.

23. Sawyer, W. W., *Prelude to mathematics*, Dover Publications, New York, 1982.

24. Tondeur, Philippe, *Vectors and Transformations in Plane Geometry*, Publish or Perish, Houston, TX, 1993.

25. Venema, Gerard A., *Foundations of Geometry*, 2nd edition, Pearson, Boston, 2012.

26. Yanik, H. Bahadir, *Prospective Middle School Mathematics Teachers Preconceptions of Geometric Translations*, Educational Studies in Mathematics, 78 (2011) 231–260.

Index